THEY CALLED ME MAD

THEY CALLED ME
MAD

Genius, Madness,
and the Scientists
Who Pushed the
Outer Limits of Knowledge

JOHN MONAHAN

BERKLEY BOOKS, NEW YORK

THE BERKLEY PUBLISHING GROUP
Published by the Penguin Group
Penguin Group (USA) Inc.
375 Hudson Street, New York, New York 10014, USA
Penguin Group (Canada), 90 Eglinton Avenue East, Suite 700, Toronto, Ontario M4P 2Y3, Canada
(a division of Pearson Penguin Canada Inc.)
Penguin Books Ltd., 80 Strand, London WC2R 0RL, England
Penguin Group Ireland, 25 St. Stephen's Green, Dublin 2, Ireland (a division of Penguin Books Ltd.)
Penguin Group (Australia), 250 Camberwell Road, Camberwell, Victoria 3124, Australia
(a division of Pearson Australia Group Pty. Ltd.)
Penguin Books India Pvt. Ltd., 11 Community Centre, Panchsheel Park, New Delhi—110 017, India
Penguin Group (NZ), 67 Apollo Drive, Rosedale, North Shore, 0632, New Zealand
(a division of Pearson New Zealand Ltd.)
Penguin Books (South Africa) (Pty.) Ltd., 24 Sturdee Avenue, Rosebank, Johannesburg 2196,
South Africa

Penguin Books Ltd., Registered Offices: 80 Strand, London WC2R 0RL, England

The publisher does not have any control over and does not assume any responsibility for author or
third-party websites or their content.

Copyright © 2010 by John Monahan
Cover design by Judith Murello
Cover illustration by Jason Seiler
Book design by Kristin del Rosario

PRINTING HISTORY
Berkley trade paperback edition / December 2010

Library of Congress Cataloging-in-Publication Data

Monahan, John.
    They called me mad : genius, madness, and the scientists who pushed the outer limits of
knowledge / John Monahan. —1st ed.
        p. cm.
    Includes bibliographical references.
    ISBN 978-0-425-23696-3
    1. Discoveries in science—History. 2. Scientists—Psychology. I. Title.
    Q180.55.D57M66 2010
    509.2'2—dc22   2010030311

PRINTED IN THE UNITED STATES OF AMERICA

10 9 8 7 6 5 4 3 2 1

To my wife,
who called me mad for wanting
to be a writer but helped me do it anyway,
to my evil creations (I mean children),
and to my dad,
who took me into the lab.

ACKNOWLEDGMENTS

There are many who made this book possible, and who deserve my undying thanks. Thank you to Margaret Miller for knowing when to nurture her little egghead and knowing when to let him go. Thanks to Joan E. Gardiner for encouraging my literary pretensions, and a very special thanks to Charles Salzberg and the rest of the New York Writers Workshop without whose assistance this may never have seen print.

CONTENTS

CONTENTS

The Real Scientists Behind the Mad Scientist

ARCANE ELECTRICAL INSTRUMENTS CRACKLE AND HUM next to bubbling glassware. The hunchbacked assistant checks the draped figure strapped to the operating table, and above it all, the genius in the long white lab coat laughs manically. These are familiar images, but where do they come from? Do real scientists spend their time constructing unstoppable giant robots, building death rays, and cloning dinosaurs? Generally speaking, no, but you might be surprised to learn that many of those mad scientists in movies and TV were inspired by actual scientists, real men and women who challenged the scientific orthodoxy and pushed the limits of intellectual frontiers.

Fictional mad scientists go back almost two hundred years. Mary Wollstonecraft Shelley created the prototype when she gave literary birth to Victor Frankenstein in her 1818 novel *Frankenstein; or, The Modern Prometheus*. While vacationing during the dreary summer of 1816, the eighteen-year-old Mary Woll-stonecraft Godwin, and her future husband, Percy Bysshe Shelley, visited the villa of Lord Byron. Cooped up by the rainy weather, they passed the time by competing to see who could compose the scariest ghost story. Shelley and Byron soon gave up, but young Mary shocked them with a tale she later said came to her in a dream. It featured an arrogant genius, who meddled in matters humans were not meant to know. He blurred the line between life and death by animating a corpse and ultimately paid the price for his hubris.

Across the Atlantic, the American Nathaniel Hawthorne added his own tales to the mad scientists' pantheon not long afterward. "Dr. Heidegger's Experiment," published in 1837, is the story of an eccentric physician who tries to find an elixir of rejuvenation. Hawthorne followed it in 1844 with "The Artist of the Beautiful," about an obsessed artist/craftsman who creates a clockwork simulacrum of a butterfly. The next year, he published "Rappaccini's Daughter," about a doctor who performs bizarre botanical experiments upon his own daughter.

By the latter part of the nineteenth century, these characters would be joined by others of the scientific fraternity. Robert Louis Stevenson's *The Strange Case of Dr. Jekyll and Mr. Hyde* thrilled the reading public in 1886. Jules Verne gave us Captain Nemo four years later in *Twenty Thousand Leagues under the Sea*. Not to be outdone, H. G. Wells made his own contributions, giving the world both *The Island of Doctor Moreau* (1896) and *The Invisible Man* (1897). Each added to the mad science mystique.

If literature created the mad scientist, the movies made him a star. From the earliest days of film, the mad scientist provided his cautionary tale. In 1910 the Edison Studios first put Frankenstein before silent-movie goers. Although it was first, it was not the movie that catapulted the doctor's creation of cobbled together corpses into the public's collective unconscious; that was the 1931 version of *Frankenstein*. When Colin Clive looked up at the heavens and proclaimed, "It's alive!" the horror movie came into its own, and the mad scientist became an indispensable part of our culture.

Countless movie incarnations followed; Rotwang, Dr. Cyclops, Dr. Morbius, the Abominable Dr. Phibes. The list goes on. After World War II, cold war paranoia and fear of the atomic bomb gave rise to the sci-fi craze, and the laboratories of the mad scientists went into production on an industrial scale. Death rays and doomsday devices competed to distract us from the actual arms race and the mad rush toward nuclear Armageddon. As they proliferated on screen, the mad scientists passed from there to the pages of comic books and the nascent world of television.

Even today, the white lab coat and the promise of forbidden knowledge, and ultimate power retain their allure. Dr. Emmett Brown went *Back to the Future*. Dr. Totenkopf met *Sky Captain and the World of Tomorrow*. Dr. Cockroach introduced us to *Monsters vs. Aliens,* and those who dare can join in with Dr. Horrible's *Sing-Along Blog*.

With all of these countless images, from the tragic to the comic, the downright evil, and the merely misunderstood, it's easy to forget that behind every great legend there is a grain of truth. Many of the elements that we take for granted—the bubbling formulas, the sparking electrical coils, the glowing lasers, and the nuclear reactors—were and are used by real scientists. The fictional mad scientist wasn't created from whole cloth. He was inspired by actual researchers who challenged convention, nurtured by our ever-increasing knowledge and the greatness of our scientific achievements and given life by our fear of going too far, too fast and the ultimate price of progress.

Here we attempt to learn more about the real men and real women, who gave rise to the myth. They are not merely caricatures and stock characters. They were actual human beings, with all the strengths and weaknesses that entails. But their vision, their passion, their willingness to push beyond accepted boundaries, their determination in the face of opposition, and their creative genius helped forge the modern age we live in. Their stories are no less amazing than their discoveries. Here then, are the lives of the real mad scientists.

CHAPTER ONE

Eureka!

The Mad Scientist Is Born

EUREKA! WHAT WORD COULD BE MORE FITTING TO START an examination of the lives of mad scientists? As it rolls off the tongue and resonates in the air, it conjures countless images of mad genius. How often has this exclamation of transcendent discovery been uttered in novels and movies and on TV? The man who etched this word indelibly into the lexicon of mad science, however, was not a fictional character. He was a man of flesh and blood, but his scientific and mathematical insights, his fantastical creations, and the story of his life gave rise to many of the mad scientist stereotypes that we hold so dear.

His genius shone like a beacon throughout the Hellenistic world, and his dazzling mathematical insights and wondrous inventions continue to fascinate us to this day. Unfortunately much of his actual life is obscured by the mists of time. In the absence of facts, a body of legend has grown, punctuated by secondhand

and thirdhand accounts of varying accuracy. Galileo venerated him. The Fields Medal, one of the most prestigious prizes for mathematicians bears his image. The tenth-century Islamic geometer Abū Sahl al-Kūhī was so impressed by his works that he called him the "imam of mathematics" (Hirshfeld, 2009). He is credited with calculating pi and the volume of the universe, discovering principles of buoyancy, inventing water pumps, and building war engines capable of grinding the Roman army to a halt. Not to mention inventing what may have been the world's first death ray. The name of this legendary genius, perhaps the greatest mathematician and inventor of all time, is Archimedes.

His life began on the sun-drenched shores of the island of Sicily, in the city-state of Syracuse. Originally a Greek colony, sitting at the nexus of Mediterranean trade, it was one of the most influential cities of the ancient world, described by Cicero as, "the greatest Greek City and the most beautiful of them all." Its harbor was filled with Egyptian, Greek, and Phoenician trading vessels bearing all manner of oils, wines, and exotic spices. Unlike most of the other city-states of the time, the leaders of Syracuse had managed to safely navigate the treacherous political waters between Rome and Carthage, the super-powers of the day. So at the time of Archimedes' birth, Syracuse enjoyed an unusual period of peace and prosperity.

Exact dates are difficult, but it is believed that Archimedes was born around 287 B.C.E. His father was the astronomer Pheidias, who passed on to the young Archimedes his love for the stars, the planets, and the other wonders of the universe. Starting around the age of seven, the boy Archimedes would have received the formal education typical of Greek males, including lessons in Greek grammar, literature, and music as well as training in sports such as running and throwing the discus and javelin.

When Archimedes was a teenager, something occurred that would have important implications for the young man. Around 270 B.C.E., Hieron, a military commander and illegitimate son of a Syracusan nobleman, seized power and became king of Syracuse. Archimedes and Hieron were friends; some have even suggested they may have been related. Whether or not that was true, the two men would form a long-lasting relationship that would serve them both well.

Shortly after Hieron took the throne, Pheidias sent the young Archimedes to continue his education across the sea in the storied city of Alexandria. Founded in Egypt near the Nile Delta by the legendary Alexander the Great, the city had been built by one of Alexander's generals, Ptolemy. When he succeeded Alexander and became king of Egypt, Ptolemy I dedicated the city to the pursuit of culture and learning, turning it into the greatest intellectual center in the ancient world.

The city was home to the temple of the Muses, the origin of our word *museum*. This wasn't one building, but a complex of buildings, including lecture halls, dissection rooms, botanical gardens, and even a zoo, in addition to accommodations for visiting scholars from all over the known world. Next to the museum was the famous Library of Alexandria. At the time, the library was the greatest repository of knowledge civilization had ever known, containing over half a million works.

At the museum, Archimedes studied with the disciples of the renowned mathematician Euclid. One of his notable teachers was Conon of Samos, not only a brilliant mathematician in his own right, but also an accomplished astronomer who studied eclipses and discovered the constellation Coma Berenices. He and Archimedes established a lifelong friendship and would frequently write letters to each other even after the latter had returned to Syracuse.

In addition to Conon, Archimedes established enduring friendships with many of his fellow students, including Eratosthenes. In later years, Eratosthenes went on to become head of the library and accurately calculate the circumference of the Earth using the length of shadows cast on the summer solstice. He and Archimedes established a long correspondence, telling each other of their latest discoveries.

While in Alexandria, Archimedes began his career as an inventor, developing a device for moving water uphill. What came to be known as Archimedes' screw consisted of a hollow cylinder. Inside of this was a central shaft around which was wrapped a long spiral blade, similar to the threads on a screw. One end of the device was placed in the water, and the other end was placed above at an angle. As the central shaft was rotated, either by hand or by draft animals, it turned the blade, which in turn pushed the water up the hollow cylinder and out the upper end. By use of Archimedes' machine farmers could easily irrigate their fields, and sailors could pump out the bilge of their ships. Modern versions of Archimedes' screw are still in use today in water-treatment plants and for moving grain; a miniaturized version is used to maintain blood flow in heart patients.

Eventually, Archimedes completed his education and returned to Syracuse, where he spent the remainder of his life. Not long after he came home, King Hieron put his newly returned friend to work. It is one of the most famous stories about Archimedes and involves a golden crown in the shape of a laurel wreath.

The crown had been given to King Hieron and was supposed to be made of pure gold, which the goldsmith had been given for the project. However, it was not uncommon for goldsmiths to adulterate or dilute gold with cheaper metals such as silver. The king was suspicious that the artist had done just that and pocketed the difference. The problem was that the crown was not only a beautiful work of art but was supposed to be used in a religious ceremony, and so was considered a holy object. If the true composition were to be discovered, it had to be done without damaging the crown itself in any way.

The king posed the question to all of his advisers, but they were unable to solve the mystery. Archimedes took up the problem. He pondered it for quite a while, but according to the story, the answer came to him in a flash of inspiration. He was visiting a public bath, and as he stepped into the tub of water, it began to overflow. The more he immersed himself in the water, the more it overflowed. He realized that the amount of water being spilled was proportional to the volume of the body being placed into the water. He further made the connection that if he placed the crown into a given amount of water, and it displaced more water than the same weight of pure gold, then the crown and the pure gold must have a different volume. In other words, if they had a different volume, then the crown could not be pure gold.

He was so excited by his discovery that he leaped from the tub and ran home through the streets, still wet and naked, yelling, "I have it! I have it!" The word for this in ancient Greek is *Eureka!* The king was delighted by the brilliance of Archimedes' insight. The goldsmith was not. Archimedes' test worked. The crown was not pure gold, and the hapless smith was executed.

When not helping out his friend the king and giving catch phrases for countless inventors and geniuses to come, Archimedes was immersing himself in mathematics. He was particularly interested in circles. Other Greek mathematicians had noted that there was a relationship between the circumference of a circle and its diameter. Today we represent this relationship with the Greek letter pi. Archimedes knew that the circumference of the circle worked out to a little more than three times its diameter, but he wished to calculate it more precisely.

First, Archimedes started by drawing a circle. He knew how to calculate the perimeter of polygons, or multisided shapes, like hexagons and octagons, so inside the circle he drew a polygon that exactly fit within it. Next he drew a polygon with the same number of sides just outside the circle so that it touched the circle within. By calculating the perimeter of both polygons, Archimedes could deduce that the circumference of the circle lay somewhere between the two.

Initially this didn't offer a very precise figure, but then Archimedes began to increase the number of sides of his polygons. He started with triangles, then went to hexagons (six-sided figures), then dodecagons (twelve-sided figures), and so on. As the number of sides increased, Archimedes came closer and closer to the actual value of pi.

Eventually, he worked his way up to a ninety-six-sided figure, an enneacontahexagon. When he was done with his calculations he wrote, "The circumference of any circle is three times the diameter and exceeds it by less than one seventh of the diameter and more than ten-seventy-oneths." In other words, he had established that pi was between 3 1/7 and 3 10/71. Archimedes didn't have the benefit of a decimal system, but if he had, his value for pi would have been between 3.1428 and 3.1408. Hitting the pi key on a modern calculator yields a value of 3.1415926, amazingly close to Archimedes' figure.

While Archimedes was wrestling with mathematical problems like this, his concentration was so intense that it may have contributed to our image of the absent-minded professor. The Greek historian Plutarch describes him as, "continually bewitched by a Siren who always accompanied him." He goes on to write that Archimedes was, "possessed by a great ecstasy and in truth a thrall to the Muses" (Hirshfeld, 2009). This possession often manifested itself in Archimedes' forgetting mundane tasks like eating or changing his clothes. He would also make use of almost any available surface to do his written calculations, including ashes from the fire, the ground outside, and even oil anointed to his own skin. Plutarch further noted that "he placed his whole affection and ambition in those purer speculations where there can be no reference to the vulgar needs of life."

Another example of Archimedes applying his powers of mathematical genius to ordinary objects involves the surface areas of spheres and cylinders. He envisioned a sphere that would just fit inside a cylinder, so that the sides of the cylinder would always be in contact with the side of the sphere. The top and bottom of the cylinder would likewise be in contact with the sphere. Archimedes then calculated the surface area of both objects, the sphere and the cylinder that contained it. He was amazed to find that the surface of the sphere was always two thirds the surface area of the cylinder. It didn't matter how large or how small they were, as long as the sphere and cylinder fit snugly together, the ratio of their surface area always worked out to 2:3. The same ratio held true when he calculated their volume. The sphere always had two thirds the volume of the cylinder. This might seem like a trivial discovery, but to Archimedes, it was wondrous, because it confirmed his belief in a universe that could be understood because it obeyed regular mathematical principles.

Math was where Archimedes took his greatest joy. He viewed his many inventions and feats of engineering brilliance as simply distractions from his mathematical accomplishments. Ironically, it was one of these distractions that lead to another famous story. Archimedes once claimed, "Give me a place to stand and I will move the Earth." King Hieron heard of the boast and told him to prove it. The king's shipbuilders had recently completed construction on what was at the time the world's largest ship. Dubbed the Syracusia, and intended as a gift to the Egyptian ruler Ptolemy, it was a monstrous three-masted vessel that weighed over two thousand tons. Now the great ship needed to be launched. The king told Archimedes that if he could move that single-handedly into the water, then he would be believed.

Archimedes quickly set to work. He constructed a complex series of ropes and pulleys and attached them to the ship. The other end of the ropes was fastened to a rotating helix, a type of large screw-like device that was attached to the dock. Word of the challenge spread, and when it came time for Archimedes to prove his boast, people from all over the city came to witness it. With the king and the populace looking on, Archimedes quietly took a seat next to the helix and began turning the handle. As the ropes grew taught, the crowd held its breath, and suddenly, the mighty ship began to move. Slowly, but steadily the genius of Syracuse pulled the hulking vessel into the water. The crowd erupted into wild cheers, and the king was so impressed that he declared, "I order, from this day on, that Archimedes is to be believed in anything he says."

His fame assured, Archimedes returned to his mathematical pursuits. Having proved he could move the world, now he set his sights on the entire universe. Specifically, he claimed that he could not only calculate the volume of the universe, but could also calculate the number of grains of sand needed to completely fill it. He started by using estimates of the Earth's diameter and the diameter of its orbit, which had been found by other mathematicians, including his friend Eratosthenes.

Next, unlike the other Greek philosophers who were using a geocentric, or Earth-centered, model of the universe, Archimedes used the heliocentric, or sun-centered, model that had been developed by a fellow Greek named Aristarchus of Samos. This was approximately eighteen hundred years before the sun-centered model of Copernicus would take center stage. Archimedes liked this model because it gave him a much larger universe in which to play his mathematical games. Based on the ratio between the Earth's diameter and its orbit, Archimedes calculated that the universe had a radius of roughly ten trillion miles. This is substantially smaller than modern estimates, but surely big enough to boggle the minds of the ancient Greeks.

Next, Archimedes further inflated the numbers he was dealing with by choosing the smallest possible size he could imagine for the sand grains needed to fill the universe. By the time he was done, Archimedes calculated that the number of sand grains needed to fill the universe was one thousand trillion trillion trillion trillion trillion grains. In our modern notation that's 10^{63}, or a one with sixty-three zeros after it. Archimedes, however, didn't have our modern notation. He was forced to work with the Greek numeral system. In that system the largest number was a myriad, equal to ten thousand, and the largest possible number, a myriad of myriad was equal to 10^8. To express his beautiful calculations, Archimedes needed to develop a new way of expressing large numbers. He presented his calculations and his new system in a work called *The Sand-Reckoner,* which Archimedes concluded with:

> These things will appear incredible to the numerous persons who have not studied mathematics; but to those who are conversant therewith and have given thought to the distance and the sizes of the earth, the sun, and the moon, and of the whole of the cosmos, the proof will carry conviction. It is for this reason that I thought it would not displease you either to consider these things.

Archimedes probably would have been quite happy to spend the remainder of his days in these sorts of abstract calculation, but alas, Syracuse's days of peace were coming to an end. The tensions between Rome and Carthage were once again building, and Syracuse was geographically as well as politically caught between the two. Sensing the winds of war, King Hieron enlisted his old friend's help in the city's defense.

During the first great conflict between Rome and Carthage in 263 B.C.E., what came to be known as the First Punic War, Hieron had formed an alliance with Rome. This left Syracuse relatively unscathed and led to a period of postwar peace and stability that lasted for decades. However, during that period, a number of pro-Carthage sympathizers in Syracuse began arguing against

THEY CALLED ME MAD   15

continuing the Roman alliance. Their argument was augmented by Hannibal's victory against the Romans in 216 B.C.E. Hieron, by then in his nineties, was feeling the pressure.

Hieron died around that time, shortly after the death of his son, Gelon, who had been serving as regent. In the power vacuum that followed, Hieron's fifteen-year-old grandson, Hieronymus, took the throne. The Greek historian Polybius described Hieronymus as, "exceedingly capricious and violent." The new king was sympathetic to Carthage and immediately switched the city's allegiance. When Rome learned of this, it would have grave consequences for Syracuse.

The Romans sent one of their leading generals to deal with Syracuse, Marcus Claudius Marcellus. He had a fearsome reputation and was a veteran of campaigns in Italy and Gaul. On his way to Syracuse, in 213 B.C.E., Marcellus and his forces captured the Sicilian city of Leontini. The Roman general ordered the beheading of two thousand of its residents.

Marcellus knew that Syracuse was well defended, but he was confident in the size of his forces and anticipated that it would take only five days to capture the city. Unfortunately for Marcellus, as Polybius wrote he, "did not reckon with the ability of Archimedes, or foresee that in some cases the genius of one man accomplishes much more than any number of hands."

The Romans attacked Syracuse by sea with a fleet of sixty quinqueremes, mighty battleships, equipped with five banks of oars, powered by 270 rowers. Each carried a crew of 30 and a force of 120 marines. Several of the ships were lashed together, carrying sambuca, a shielded ladder used for scaling defending walls. Marcellus knew that Syracuse was protected by long-range catapults, but he counted on his fleet's speed to quickly carry them inside the defender's range. They could then close on the city while the catapults overshot them. Archimedes anticipated this. He built banks of catapults with successively shorter ranges. These were able to rain death and destruction down upon the invaders during their entire inward passage.

According to legend, Archimedes then let loose with his greatest weapon, the world's first death ray. It consisted of a series of highly polished shields arranged as a parabolic mirror to focus the sunlight on incoming ships and set them ablaze. The claims about these burning mirrors were first made by the second-century C.E. historian Lucian and have been debated and tested by everyone from René Descartes and students at M.I.T. to the TV show *MythBusters*. Attempts to re-create the device have met with mixed results, but the legend lives on. Even without the ability to make ships burst into flame, the reflecting mirrors would have interfered with the vision of archers on the invading ships and made it difficult, if not impossible to return fire. The fleet was forced to retreat.

To avoid a repeat, Marcellus ordered a nighttime attack. Under cover of darkness, his ships were able to reach the defending city's wall, but as soon as they approached, they came under a withering hail of arrows and darts. Archimedes had installed a series of loopholes in the walls. From behind these holes the defenders could fire crossbows from the safety of cover.

Meanwhile, Archimedes used cargo cranes to swing out over the walls and drop boulders on the Roman ships. According to an account by the historian Plutarch:

There was discharged a piece of rock of ten talents weight [about six hundred pounds], then a second and a third, which, striking upon it with immense force and a noise like thunder, broke all its foundation to pieces, shook out all its fastenings, and completely dislodged it from the bridge.

At this point, from out of the darkness, the claw of Archimedes descended on the invaders. It was another modified cargo crane, essentially a giant lever, with a heavy counterweight. The Syracusians dropped grappling hooks from the crane to snag the Roman ships, and then lifted them out of the water, shaking them to bits and dashing their crews on the nearby rocks. Marcellus was once more forced to order a retreat.

Having been twice routed in his naval assault, Marcellus ordered a land attack. Once again, Archimedes unleashed his engines of destruction on the invaders as they marched toward the walls. Plutarch described the effect that these inventions had on the morale of the Roman soldiers:

> Such terror had seized upon the Romans, that, if they did but see a little rope or a piece of wood from the wall, instantly crying out that there it was again, Archimedes was about to let fly some engine at them, they turned their backs and fled (Hirshfeld, 2009).

What the great Roman general Marcellus had once confidently predicted would be a five-day victory, eventually dragged out into a two-year siege. Archimedes' brilliance had stymied them at every turn, but eventually Roman perseverance paid off. One night, during a festival when some of the defenders of Syracuse had partaken of too much wine, the Romans managed to find one section of the wall unprotected. The invaders quickly took advantage and scaled the wall with a few soldiers who managed to get inside and open one of the gates.

When the end came, the death of Archimedes would do as much to inspire the legend of the mad scientist as his life had. As Roman soldiers stormed the city, Archimedes, as usual, was drawing circles and making elaborate calculations in the sand outside of his home. Recognizing the value of such a man, Marcellus had given specific orders that Archimedes was to be captured unharmed, but when a Roman soldier confronted him, instead of fleeing or surrendering, Archimedes simply yelled at him, "Don't disturb my circles." The angry soldier, not recognizing one of the greatest geniuses who ever lived, promptly killed him.

Marcellus was reportedly so angered by the loss, that he ordered the offending Roman soldier executed. The general who had once casually beheaded two thousand of his opponents at Leontini, now showed his respect for the brilliant defender of Syracuse by holding a grand funeral in Archimedes' honor. They gave him the most fitting of tributes. His grave was marked, as Archimedes had wanted, with a stone tablet engraved with a diagram showing a sphere inside a cylinder. The epitaph for the most influential inventor and mathematician in history proclaimed the perfect 2:3 ratio between the two.

Much of Archimedes' legend was shaped by his time in Alexandria. Within the halls of the city's great museum, he studied with some of the finest minds the ancient world had to offer. For over seven centuries the Temple of the Muses would play host to the likes of the astronomer Aristarchus; the great physician Galen; and the father of algebra, Diophantus, among others. The greatness of Alexandria, however, like all good things, eventually came to an end, but in its death throes, it would make one last contribution to the mad scientist legend.

By the fourth century, the city had lost much of its former glory. Roman rule had sapped much of its wealth, and it was wracked by civil and religious discord. Most of the great library had burned. The museum had fallen into disrepair, and its last recorded member was a philosopher and mathematician named Theon. Although he was a brilliant philosopher and teacher in his own right, who made accurate predictions of both solar and lunar eclipses, his major claim to fame comes from being the father of a remarkable young woman. Renowned for her beauty and wisdom, her mathematical and philosophical accomplishments would enshrine her among Alexandria's most luminous residents. Her teachings would go on to influence some of the greatest Pagan and Christian leaders. Her brilliant life and tragic death would mark both high and low points in the history of the ancient world. Her name was Hypatia of Alexandria.

Hypatia was born sometime around the year 355 C.E. The exact date is unknown, but in a time when girls were rarely educated, and women were confined to proscribed roles, Theon gave his daughter the benefits of a first-rate education. He tutored Hypatia in the intricacies of mathematics, astronomy, and philosophy, and she proved herself a gifted student. The early Christian historian Socrates Scholasticus described how, "She inherited her father's extraordinarily distinguished nature, and was not satisfied with the training in mathematics that she received from her father, but turned to other learning also in a distinguished way."

Soon young Hypatia was not only learning from Theon, but helping him with his own work as well. As part of his duties at the museum, Theon edited and wrote commentaries on earlier texts, including Euclid's *Elements* and the works of the Greek astronomer Ptolemy, not to be confused with the Egyptian king. Hypatia helped her father, and Theon's commentary on the *Almagest* includes the following introduction: "Commentary by Theon of Alexandria on Book III of Ptolemy's *Almagest*, edition revised by my daughter Hypatia the Philosopher." His student had become his collaborator.

Within a short time, Hypatia was writing her own commentaries, including works on Apollonius of Perge, the great geometer, and *Arithmetica* by Diophantus. Her writings on Diophantus are of particular note because, although he is considered the father of algebra, he was a notoriously dense and difficult to understand writer. Without Hypatia's clear and concise explanations, his original work might not have survived or gone on to form the basis for modern mathematics.

Not satisfied with simply commenting on the discoveries of others, Hypatia was soon undertaking her own research. In the course of doing so, she developed a new type of astrolabe, a device used by astronomers to measure the position of the sun and stars. She also developed a number of tools for making more earthly observations, including a hydroscope for looking at objects under water and a hydrometer, used to measure the density or specific gravity of liquids.

By around 400 C.E., Hypatia had succeeded her father as head of the neoplatonist school of philosophy. Students flocked to her, and she soon counted among them the sons of some of the most powerful families in the Roman Empire. Unlike many schools of the time, Hypatia's school freely admitted Christians,

Pagans, and Jews, although she herself was Pagan. Her students soon formed a loyal core that referred to each other as "brother."

Unfortunately, none of Hypatia's own writings still exist, but much about her can be gathered from letters written to and about her by her loyal students and accounts written by others during her life. Again, Socrates Scholasticus described her in his *Ecclesiastical History:*

> There was a woman at Alexandria named Hypatia, daughter of the philosopher Theon, who made such attainments in literature and science, as to far surpass all the philosophers of her own time. Having succeeded to the school of Plato and Plotinus, she explained the principles of philosophy to her auditors, many of whom came from a distance to receive her instructions. On account of the self-possession and ease of manner, which she had acquired in consequence of the cultivation of her mind, she not infrequently appeared in public in presence of the magistrates. Neither did she feel abashed in going to an assembly of men. For all men on account of her extraordinary dignity and virtue admired her the more.

At a time when proper Alexandrian women were rarely seen outside the home, Hypatia would frequently deliver public lectures in the city center. Dressed in the rough-textured white *tribon* robes worn by her male counterparts, she would speak with great eloquence on mathematics and astronomy or the history and philosophy of Plato and Aristotle. The crowds gathered to hear her speak were captivated by the breadth of her intellect, the passion of her words, and the extent of her beauty. No paintings or statues of Hypatia survive today, but her physical presence was so well known that the nineteenth-century French poet Charles Leconte de Lisle described her in one of his poems as having: "The spirit of Plato and the body of Aphrodite."

In spite of, or perhaps because of this, Hypatia reportedly remained celibate, and dedicated her life to science, philosophy, and teaching. In one famous story about her, Hypatia was delivering a lecture to her students on the nature of beauty. As she told them:

> When a man sees the beauty in a woman's body, he must not seek to conquer her with his lust, but realize instead that her beauty is but an image of what real beauty is. By sinking to the lowest depths of his animal nature, he is not contemplating the true essence of beauty, but in his blindness is actually consorting with the illusive shadows of Hades (Vrettos, 2001).

Apparently one of her students failed to get the message, because after the lecture, he came up to her and confessed that he loved her. To further make her point, she reached into her bag and took out the Egyptian equivalent of a menstrual pad, telling the young man, "This is what you love. You do not really love beauty."

This idea, that behind every object there lay a deeper reality, an essence of the true object, was at the core of Hypatia's and the other neoplatonists' philosophy. Like Plato, they believed in the power of rational thought. Hypatia put great value in Plato's technique of questioning to attain greater understanding. Even though she was a Pagan, she cautioned against blindly believing in myth. In her own words, "Fables should be taught as fables, myths as myths, and miracles as poetic fancies. To teach superstitions as truth is a most terrible thing. The mind of a child accepts them and only through great pain, perhaps even tragedy, can the child be relieved of them" (Donovan, 2008). Those words would become prophetic in the years to come.

As Hypatia's fame spread, her opinions came to be valued by the city's leaders. They bestowed on her many civic honors and frequently the chariots of prominent citizens, councilors, and even military commanders could be seen coming and going from her home as they sought her advice on important matters of the day. Even Orestes, Alexandria's imperial prefect appointed by the emperor in Rome, sought her council. The two became close friends, and they were frequently seen strolling through the streets of the city, discussing issues both philosophical and political.

Admittedly, a beautiful, articulate, and influential woman, no matter how great her genius or how impressive her inventions, does not quite meet the popular image of the mad scientist. Tragically, it is her death, perhaps more so than her life, that links her inextricably with that image. Hypatia, the most gifted of mathematicians, astronomers, and inventors; beloved teacher; and trusted council of the rich and powerful, met her end, like many a fictional mad scientist, at the hands of an angry mob.

By the early fifth century, Alexandria was caught in the midst of a major power struggle between Orestes, head of the civil authority, and Cyril, the Christian patriarch of the city. Even though she was Pagan, Hypatia had managed to steer clear of many of the antipagan purges that took place as the Roman Empire became increasingly Christianized. As was mentioned earlier, she admitted students from all religions into her school, a number of whom became prominent leaders in the church. But, as the struggle between church and state intensified, Hypatia found herself caught in the middle.

Cyril had learned early on that one of his greatest tools in his rise to power was the power of the mob. By giving them something to rally against, he was able to harness them for his own purposes. That's how he achieved his leadership position within the city's church.

When Theophilus, the former patriarch of Alexandria, died in the year 412 C.E., two candidates were in a position to succeed him. Timothy, his archdeacon, had the support of the church hierarchy. Cyril, Theophilus' nephew, had the support of his followers. After three days of savage street fighting, Cyril was installed as the new patriarch. He went on to cement his power by directing his followers against fellow Christians who were members of the Novitian sect. After that, he turned them on the city's large Jewish population.

Rumors spread wildly and mob violence ensued. The pagan temple Serapium was sacked and its library burned. Cyril sent agent provocateurs to provoke the Jewish community, and Jewish radicals within that community ambushed Christians. For months the attacks and counterattacks raged, and the civil authorities seemed helpless to stop it. The streets ran red with Pagan, Jewish, and Christian blood.

Faced with the increasing violence, Hypatia threw her considerable support behind Orestes. Cyril, who had long been jealous of Hypatia's popularity and influence, responded by spreading rumors that she was practicing black magic. He accused her of being a sorceress who was, "devoted at all times to magic, astrolabes and instruments of music" (Reid, 2006). As ridiculous as the claims were, they had the desired effect on Cyril's followers.

In 415 C.E. a Christian mob attacked Hypatia. They confronted her chariot one evening while she was on her way home. The enraged zealots grabbed her from the vehicle, ripped off her philosopher's robes and dragged her naked through the street to a nearby church. It had once been the Caesareum, a temple built by Cleopatra for her beloved Mark Antony, but had since been converted to the empire's new religion. There, inside the Christian church the mob threw Hypatia to the floor of the nave. They beat and stabbed her to death with pieces of broken tile and pottery. Once she was dead, the crowd tore her body apart and carried the mutilated corpse outside the city walls where it was burned on a bonfire.

It was a brutal and tragic death for a woman who may have been one of the most brilliant minds of her time. There are many historians who use Hypatia's death to mark the end of the classic period, the death knell for a golden age of reason. In the aftermath of her murder, the Christian church compounded the crime by destroying her written works in a futile attempt to cover up the incident. Today, none of her original works survive, and we have only historical accounts and letters from her students to testify to her genius. Hypatia's greatest legacy may be the words that she passed on to her students. She told them, "Reserve your right to think, for even to think wrongly is better than not to think at all." With those words, the legend of the mad scientist would live on.

CHAPTER TWO

Sulfur and Brimstone

The Alchemist's Dream

EARLY IN THE NOVEL *FRANKENSTEIN; OR, THE MODERN Prometheus,* Mary Shelley gives some tantalizing clues to the inspiration for her fictional mad scientist, Victor Frankenstein. In the story, he recalls an incident from his youth where he comes upon the medieval writing of Cornelius Agrippa. Victor's father is dismissive, calling it, "sad trash," but the younger Frankenstein is fascinated. He pours through the work, and when he is done, relishes the thought of obtaining more, "When I returned home my first care was to procure the whole works of this author, and afterwards of Paracelsus and Albertus Magnus. I read and studied the wild fancies of these writers with delight; they appeared to me treasures known to few besides myself" (Shelley, 1819).

Who were these men, who so captivated the young Victor? Were Cornelius Agrippa, Paracelsus, and Albertus Magnus simply additional figments of Shelly's imagination? No, these were actual historical figures. However, these writers, who played so seminal a role in the fictional life of Frankenstein, were not modern men of science. They were from a much older pedigree. They were, in fact, alchemists.

The collapse of the Roman Empire and the final destruction of Alexandria may have marked the end of the Ancient world, but much of its knowledge lived on. The Arabs preserved and expanded upon it, translating it from Greek into Arabic as they went. In particular, they took many of the Alexandrian secrets of metalworking and gave them the name *alkimia*. That body of knowledge gave rise to what we now call alchemy.

For centuries, while Europe dissolved into the warfare and chaos of the Middle Ages, the great Arab alchemists practiced their craft. They carefully melded the knowledge of the Greeks and Egyptians with their own and other secrets from farther east into an amalgam that they jealously guarded. They then recorded their discoveries in encrypted journals to prevent others from learning their secrets.

Eventually, however, like all good secrets, word of their work spread. By the eleventh century, alchemical texts started to appear in Europe. Painstakingly, they were translated from Arabic into Latin and gave birth to European alchemy. This school borrowed many concepts from its Middle Eastern predecessors, including the philosopher's stone and the elixir vitae. The former was a mystical substance that could transmute lead into gold, and the latter was a potion capable of curing any illness and granting immortality.

As they toiled in secret over arcane glassware and crucibles to unlock the secrets of limitless wealth and extended life, it's easy to see how the alchemists would inspire our modern image of the

mad scientist. One in particular, however, mentioned specifically
in Mary Shelley's work, would come to engender the image more
so than others. There seemed to be no limit to the fantastical
claims made about him. It was said that he had mastered the
secrets of immortality and could raise the dead. Others reported
that he could speak with spirits, and rode a white horse given to
him by the devil. More than three hundred years after his death,
people continued to try to invoke his healing powers by making
pilgrimages to his grave in Salzburg. The real story of his life is
hardly less impressive. Although he was born Philippus Aureolus
Theophrastus Bombastus von Hohenheim, he would come to be
widely known by his self-appointed name, Paracelsus.

In the late 1490s, an itinerant physician named Wilhelm von Hohenheim, the bastard son of a German nobleman, settled in the
rustic Swiss village of Einsienden. Nestled on the wooded moun-
tainside near the Siehl River, it consisted of little more than an inn
and a scattering of houses overlooked by the local abbey. This was
where the doctor settled and took a local woman, Elsa Ochsner,
as his wife. In 1493, they had their only child, a son. To reflect his
father's love of learning and classical philosophy, they named him
Theophrastus after the man who succeeded Aristotle as head of
the Lyceum in Athens. To this they added Philip, the name of the
Christian saint on whose day the child was born.

The young Philip Theophrastus suffered a difficult childhood.
Despite his father's being a doctor, the family was often im-
poverished, and the poor diet the boy had to endure resulted in
rickets. This led to skeletal deformities that handicapped the

child. Despite this, he seemed oddly proud of his rough upbringing. He later wrote:

> By nature I am not subtly spun, nor is it the custom of my native land to accomplish anything by spinning silk. Nor are we raised on figs, nor on mead, nor on wheaten bread, but on cheese, milk and oatcakes, which cannot give one a subtle disposition. Moreover, a man clings all his days to what he received in his youth; and my youth was coarse as compared to that of the subtle pampered, and over-refined. For those who are raised in soft clothes and in women's apartments and we who are brought up among the pine-cones have trouble in understanding one another well. To begin with, I thank God that I was born a German, and praise Him for having made me suffer poverty and hunger in my youth (Ball, 2006).

Hunger was far from the only danger the young man faced. Many retellings of his life include reports that he was castrated in his youth. Some say that this was a grievous accident caused by an encounter with a wild boar. Others speculate that the emasculation took place at the hands of drunken soldiers. Whether either account is true is up for some debate, but it is known that for whatever reason, Philip Theophrastus never married or sired any children. He apparently remained celibate for the remainder of his life.

As if this were not already the type of upbringing capable of creating a fictional mad genius, when Philip Theophrastus was only nine years old, his mother died. Death was a common companion in the medieval world, but the death of Philip Theophrastus' mother seems particularly tragic. According to several accounts, Elsa suffered from some sort of mental illness, perhaps what today we would call depression or bipolar disorder. Regardless of the specific diagnosis, when her son was nine, Elsa walked to the nearby Teufelsbrücke, or Devil's Bridge; climbed the parapet; and flung herself to her death in the raging waters of the Siehl River.

Upon the death of his mother, his father moved them from the isolated Swiss village to the mining city of Villach in what is today Austria. This was a thriving center for mining and trade at the time, controlled by the prosperous Fugger family. Although he was primarily a physician, Wilhelm knew enough about mineralogy to get a job teaching at one of the local mining schools.

He enrolled his son in a nearby Benedictine monastery school. Philip Theophrastus received a classical education there, typical of the day, including Latin, grammar, logic, and rhetoric. Perhaps more significant though, he made the acquaintance of one of the teachers, Bishop Erhart, who was, among other things, an accomplished alchemist who would give his new student a thorough introduction to the ancient art.

He took to this topic with apparent relish, but at the age of fourteen, decided to follow in his father's footsteps and become a physician. With this goal in mind, he left home and took to the open road in search of further education. This was not uncommon at the time. Students would frequently wander from university to university seeking the best teachers. Although it was common, this was not an easy way of life. Traveling students had to live by their wits to navigate the dangers of the road and would frequently beg for their meals or earn a modest living by singing in inns, pulling teeth, and selling minor cures. In the course of two years, he traveled to universities in Heidelberg, Leipzig, Wittenberg, Cologne, and Tübingen.

At each of these schools, he received the standard medical education of the day, which relied almost exclusively on the classical texts of the ancient Greeks, like Galen and Hippocrates. In addition, doctors of the day were trained to distance themselves from their patients. Being a physician was seen as holding a position of high status. Doctors were entitled to ride fine horses, wear fur hats and red robes, and even marry into noble families. Those of such status did not sully themselves associating with, let alone physically treating, the rabble. That was seen as the work of the lowly surgeons and barbers. In the schools that educated such men, brute memorization and recitation of the standard texts would have taken the place of any practical experience.

The young Philip Theophrastus chafed under such a stifling environment, and it's unclear if, during all of his travels, he ever actually received a degree. He claimed that he received his baccalaureate in medicine from the University of Vienna in 1510, and later claimed to have been awarded a doctorate from the University of Ferrara. In both cases, however, university records are missing or impossible to verify.

By this point, he was already starting to display the type of strong opinions typical of a mad genius. When he was only nineteen, he declared, "At all the German schools you cannot learn as much as at the Frankfurt Fair." If that weren't enough to earn the displeasure of his teachers, he later was quoted as saying, "All the universities and all the ancient writers put together have less talent than my ass" (Ball, 2006). He was never one to mince words.

It was at about this time that he decided to adopt a new name. Celsus was a second-century Greek writer and author of one of the most prominent medical texts used in the Middle Ages. In an obvious attempt to show his disdain for such ancient wisdoms, the would-be doctor took the name Paracelsus, meaning "equal to Celsus."

Having completed his education, the newly dubbed Paracelsus could have been expected to settle down into a comfortable local practice, but settling down was not in his nature. He was soon wandering again. Over the course of the next several years, he wandered through almost every country in Europe, and many beyond. What's more, while he traveled he took advantage of every opportunity that presented itself to learn, not from the dry academics who had chained themselves to the past but from those who possessed the practical knowledge of healing. As he put it, "A doctor must seek out old wives, gypsies, sorcerers, wandering tribes, old robbers, and such outlaws and take lessons from them" (Ball, 2006). This quest for knowledge took him to the farthest reaches of Europe, and as he went, he put his newfound knowledge to use, treating the wealthy for a fee and providing his services free of charge to the poor.

In the course of his travels, he became an army surgeon. First, he joined up with the army of Dutch insurgents, who were in open revolt against Charles III of Spain. From there, he served with the Danes when they invaded Sweden. Eventually, he made his way into the service of the legendary Teutonic Knights.

At around this time, Grand Duke Vasily III of Moscow was trying to consolidate power. His father, Peter the Great, had declared himself czar of all the Russias, but Russia had suffered the depredations of the Mongols for centuries and was physically, spiritually, and intellectually isolated from the rest of Europe.

Vasily vowed to change things, and toward that end invited Western physicians, astrologers, architects, and humanists to visit his court in Moscow. Not being enamored of the type of Teutonic discipline being enforced on him at the time, Paracelsus leaped at the chance.

He traveled to Moscow by sledge in the winter of 1521. Unfortunately, shortly after Paracelsus' arrival, a local Tartar warlord decided to invade the city. No sooner had the young physician arrived than a force of a hundred thousand Tartar warriors laid siege to the capital. Ultimately, the invasion was thwarted, but in the chaos that ensued, Paracelsus and a group of fellow Western visitors were captured.

Fortunately for Paracelsus, the Tartars held healers in high regard, akin to holy men. They took him back to their home in the Crimea, and there he was able to ingratiate himself to his captors by once again displaying his medical skills. He also took advantage of the situation by learning whatever he could from the wise men and shamans of these people. He even claimed that one of the Tartar leaders was so impressed that he asked Paracelsus to accompany him on a diplomatic mission to the legendary city of Byzantium, and that once there, he met with Arab alchemists who revealed to him the secrets of the philosopher's stone. Like many things in Paracelsus' life, this too is impossible to verify.

His return from the Tartars was facilitated by a group of wandering Polish knights, who "rescued" him and returned him and the other captives once more to Europe proper. Many would have taken this as a cue to go home, but Paracelsus continued his travels. He made his way to Venice and once more entered the service of the army as a surgeon. In the course of his service, he traveled along the Venetian trade routes to Crete and even to Alexandria itself, continually picking up what knowledge he could along the way.

Eventually, however, even the vagabond Paracelsus tired of the road, and in 1524, he returned to his father's home in Villach. He brought back with him a large broad sword, given to him while serving in the Venetian army, and a new drug he claimed to have discovered, called laudanum. Despite the fact that no one ever saw him actually use the sword, he kept it with him at all times, even when he slept. This led to persistent rumors that he kept a supply of the new drug, likely an opium derivative, secreted away in the sword's pommel.

Rested and refreshed, Paracelsus set off to establish a medical practice in the nearby city of Salzburg. Unfortunately, even when he intended to settle down, he could not stay out of trouble. Shortly after he got there, there was a peasant uprising, and he made the mistake of following his antiestablishment instincts, and associating himself with the rebels. When they were put down, he was forced to flee. He ended up in Strasbourg.

Strasbourg was unusual at the time, because surgeons there were considered to be equal in status to physicians. It was home to one of the finest surgical schools in all of Europe, and Paracelsus sought a teaching position. By this point, however, his travels and unconventional medical practices had earned him a certain notoriety, and many of the faculty did not welcome such a disruptive personality into their midst. One of the school's surgeons, Wendolin Hock, challenged Paracelsus to a public debate. Scholarly debates, such as this, were based more on eloquence and debating skill than on practical knowledge. This put Paracelsus at a distinct disadvantage since he reportedly stuttered. What's more, the chosen topic for the debate was anatomy. Not his strong suit. The affair was a humiliating defeat.

His humiliation was assuaged, however, by the arrival of a summons for his medical expertise. The wealthy and well-connected humanist publisher Johann Froben in the city of Basel, seventy miles to the south, was suffering from a badly infected leg. His pain was so great and the prognosis so grim, that all of the local physicians recommended amputation. This was several centuries before the advent of effective anesthesia, and the chances of surviving such an operation were not good. In desperation, Froben remembered rumors that he had heard of a new doctor who was making a name for himself with supposedly miraculous cures. He felt that his only hope lay in summoning the young physician, and Paracelsus heeded the call.

Paracelsus moved into the publisher's house, and began administering his newfangled treatments. To Paracelsus' credit, and Froben's relief, it worked. The leg was cured. Coincidentally, the influential scholar, Erasmus of Rotterdam, was also staying in the publisher's home at the time. Having seen the physician's great success with his friend, Erasmus asked if Paracelsus might not be able to treat him for gout and a kidney disorder he was suffering. Again, the unconventional doctor effected a cure. His fame soared.

Paracelsus, with the patronage of his two newly impressed patients, managed to secure for himself an appointment as municipal physician and professor of medicine at the University of Basel. Again, the local medical establishment would have none of

it. They challenged the appointment and demanded that a col-
loquy, a public test, be held to determine the young firebrand's
qualifications. What's more, the physician they chose to admin-
ister the test was none other than Paracelsus' nemesis from Strasbourg, Dr. Hock.

Paracelsus was once burned, twice cautious about another
encounter with Hock. He and his allies countered that, since the
position had been created by the council, it did not require the
approval of the university. The university officials responded that
since they weren't required to give their consent, then they
weren't required to give Paracelsus university privileges either,
even a lecture hall in which to teach.

This squabbling continued until the start of the summer semester, at which point Paracelsus announced that he would not
only lecture on his new brand of medicine but would do so in
German, rather than the traditional Latin, to make this teachings
more accessible. As he put it, "What a doctor needs is not eloquence or knowledge of language and books, but a profound
knowledge of Nature and her works" (Ball, 2006). What's more,
since the lecture would have to be held in a private hall off campus, he took the liberty of opening the lecture to anyone who
cared to attend, including academics, physicians, surgeons, barbers, alchemists, and even lay people.

Word of all this quickly spread, and on the appointed day,
the hall was packed with students, townspeople, and local physicians curious to see what sort of heresies and nonsense this
newcomer would spout. Paracelsus entered dramatically in full
professorial robes. He then proceeded to shuck them off to reveal
the well-worn alchemist apron he had on beneath. He announced
that he would reveal to those assembled medicine's greatest se-

cret. At which point, he uncovered for his audience, with great flourish, a fresh bowl of steaming human excrement.

The reaction of the audience was immediate and unrestrained. As the hall exploded into a mixture of horrified gasps, unrestrained laughter, and all manner of other exclamations, the school's assembled faculty stormed out. As they did, Paracelsus called after them, "If you will not hear the mysteries of putrefactive fermentation, you are unworthy of the name of physicians" (Ball, 2006).

This was an instant hit among the students, and for the remainder of the semester, they faithfully attended, while Paracelsus expounded on his theories of medicine. He denounced the theory of the four humors advocated by most doctors at the time. Instead, he maintained that most bodily functions were the result of alchemical processes being conducted within the body and that therefore most diseases could be treated with chemical preparations. This laid the foundation for something called iatrochemistry, a branch of alchemy that dealt with the chemistry of living things, what we might today call chemotherapy.

It is important to remember, however, that despite this, Paracelsus was far more an alchemist than a chemist. He had his own ideas, just as strange and unfounded as the four humors. For instance, he claimed that one could determine the therapeutic value of a plant by looking at its shape. Liverwort resembles the human liver and therefore could be used to treat disorders of that organ, and orchids, whose flowers are shaped vaguely like testicles, could be used to restore a man's virility.

He also had some unique views on the composition of matter. He distanced himself from the Aristotelian idea that everything was made up of four elements, fire, earth, air, and water. But his own views were a far cry from modern chemists. He claimed that all matter, including that which made up living things, was made up of three principles: mercury, sulfur, and salt. These weren't the mercury, sulfur, and salt that we might be familiar with. They were closer to some sort of essential principle and each conferred specific properties to matter. For instance, the principle of mercury imparted luster to things. Sulfur gave combustibility, and salt gave salinity.

However, aside from these eccentricities, Paracelsus did much to further medicine. He railed against the divisions between physicians and surgeons, referring to physicians who relied on theories instead of practical knowledge as "high asses," and disparaging the surgeon who lacked any theoretical knowledge as a "wood doctor and fool." Paracelsus incorporated much of this into two of his medical texts *Paragranum* and *Opus Paramirum*.

Three weeks after his initial lecture, on June 24, Paracelsus added insult to injury. It was St. John's day, traditionally a time when students would drink and make merry. They would often build a large bonfire in the center of the market and hurl into it things that they felt they no longer needed. Caught up in the excitement, Paracelsus flung a copy of a book by Galen into the flames. Again, the students cheered. The faculty did not.

No matter how much he irritated them though, because he had the backing of Froben and the council, there was little that the university officials could do. That all ended when Froben decided to attend the Frankfurt Fair. It was four hundred miles away, and Paracelsus warned him against the arduous journey, but the publisher ignored the advice. While there, he suffered a stroke and died. Paracelsus' enemies leaped on the opportunity and began spreading rumors that the death had been caused by the doctor's unconventional medical treatments. At the urging of the university, an investigation was launched.

Paracelsus might have survived the inquiries, but at about the same time, one of the richest men in the city, Cornelius von Leichtenfels fell gravely ill. He offered the unheard of sum of a hundred guilders to any doctor who could cure him. Paracelsus accepted the challenge and quickly set to work. He was perhaps too quick, because once he had cured Leichtenfels, the wealthy patient decided that six guilders was a more appropriate fee for the amount of work done.

Paracelsus was outraged about being cheated out of his fee, and he made the impolitic decision to sue. The local magistrates, not surprisingly, sided with Liechtenfels. In retaliation, Paracelsus wrote a pamphlet denouncing them, and even though it was published anonymously, there was no mystery about who had written it. Libeling a magistrate was a punishable offense and Paracelsus was forced to flee the city to avoid arrest.

News of the Basel scandal followed Paracelsus wherever he went. The authorities in Colmar forced him to leave, and those in Esslingen followed suit. Eventually, he made his way to Nuremberg. The local physicians there challenged him to another public debate. Paracelsus declined the offer and proposed instead that they provide him with a patient that they considered incurable, and he would cure him. Confident that he could do no such thing, the physicians directed him to a nearby hospital for quarantining lepers.

Back then, leprosy was not so much a specific diagnosis as it was a general category used for any number of wasting disorders afflicting the skin. That included one of the newest plagues to attack Europe, syphilis. This gave Paracelsus just the chance he needed. Having seen it frequently in the course of his travels, he was well acquainted with syphilis, and it provided the perfect opportunity to put his alchemical knowledge to work.

He treated the patients with the alchemist's old standby, mercury. He wasn't the first to use mercury to treat skin disorders. The Arabs had been doing it for quite some time. The problem was that mercury is a deadly poison. Those who even breathe the fumes experience severe swelling of the throat and mouth. Their gums turn black. Their teeth fall out. They can experience tremors, convulsions, and death, a clear case of the cure being worse than the disease.

Paracelsus' insight was in realizing that it wasn't the poison as much as the size of the dose that was responsible. As he said, "Poison is in everything, and no thing is without poison. The dosage makes it either a poison or a remedy." He was able to treat the patients with very small doses, alleviating their symptoms, while avoiding the worst of the side effects. Of the fourteen supposed lepers that he was presented with, he successfully treated nine of them. In fact, mercury-based medicines would remain the standard of treatment for syphilis until the early twentieth century. Because of his understanding of the relationship between dose size and toxicity, Paracelsus is often credited as being the father of toxicology.

The local physicians where not swayed. What's worse, Paracelsus' mercury treatment was considered a direct threat to the conventional treatment, an imported wood from the New World called guaiac. Realizing the superiority of his mercury treatments, Paracelsus published a pamphlet recommending against the use of the imported wood. It was just his luck that one of the most influential merchant families in the city, the Fuggers again, had a monopoly on the importation of guaiac and didn't want anything to hurt their business. When Paracelsus tried to publish a book titled *Essay on the French Disease*, they pressured local officials to censor it.

The next several years form a curious period in Paracelsus' life. As he crisscrossed Europe once again, his attention seemed to turn more to religion than healing. This was a time of great religious upheaval. Martin Luther and his disciples were denouncing the abuses of the Catholic Church, which in turn, responded by cracking down on dissent. The Anabaptists were in open and often violent rebellion, while the humanists tried in vain to argue for some sort of moderation.

In the midst of all this, Paracelsus continued to tend to the medical needs of the poor but also sought to minister to their spiritual needs as well. He lacked a pulpit but made up for it by preaching to them in the street and in taverns, wherever he could find an audience. He lauded the spiritual benefits of poverty, excoriating the rich, and lived himself by begging. He said, "Blessed and more blessed is the man to whom God gives the grace of poverty." He even refused to wear shoes, carrying his message barefoot to the common man. However, he never lost his talent for abusing the powerful and making enemies. At one point he denounced both Martin Luther and the pope as, "two whores discussing chastity."

His fortunes changed, as they had so many times in the past, when he wandered into the northern Italian city of Merino. He began working on a book he modestly titled *The Great Surgery Book*. This was published in 1536 and focused less on how to perform surgery than on ways to avoid it. Given the state of the surgeons craft at the time, the book should therefore be credited with saving many lives. It was an immediate success and prompted a second printing only a year later. Between the success of his book and his growing list of patients, Paracelsus was soon living a comfortable life.

In 1540, Paracelsus found a place to settle down. Duke Ernst of Bavaria, the prince-bishop of Salzburg offered him a permanent appointment. After wandering the far corners of Christendom and beyond, he at last had a home, however briefly. In the fall of 1541, a few months before his forty-ninth birthday, Paracelsus died. There are conflicting reports about the cause of death. According to one, he was beaten to death by thugs hired by his enemies. Another account claims that he simply fell while drunk. There are even those who speculate that he died from drinking one of his own potions.

Regardless of the cause, three days before he died, Paracelsus made a will. It specified which psalms were to be sung at his funeral and made provisions for a coin to be given to every poor man who attended. He left a few guilders to a relative and his executors and bequeathed his instruments and medicines to the local barbers.

It is difficult to imagine a figure that could better inspire our image of a mad scientist than Paracelsus. He was an uncompromising iconoclast, a scorned outsider, a man whose ideas were years, if not centuries ahead of their time. While many of his alchemical beliefs are clearly mistaken, he can be credited with being an ardent experimentalist during a period when questioning the wisdom of ancient texts was literally considered heresy. Two centuries after Paracelsus' death, the influential French chemist Antoine-François Fourcroy published a history of chemistry in which he divided the early history of chemistry into four epochs: the ancient Egyptians, the Arabs, the alchemists, and the medical or pharmaceutical chemistry of Paracelsus.

·☼·

Scholars frequently differentiate between the alchemists, like Paracelsus, and the scientists that came later, but the line between the two is not as clear cut as some would like to imagine. It was a gradual transition, and for a while the two existed side by side and sometimes were practiced by the same individuals. Take for example a man whose name symbolizes the very epitome of a scientist in the minds of many. How shocked they would be to learn that the man, who is often cited as one of the fathers of modern science, was also a practitioner of the older arts. While he revolutionized the nature of mathematics and developed scientific theories that would stand for centuries, he also chased the ancient dream of the philosopher's stone. The name of the man who so gloriously embodied both science and magic, and in doing so contributed mightily to the legend of the mad scientist, was none other than Isaac Newton.

Hannah Ayscough Newton was the recent widow of a local yeoman farmer in the rural English village of Woolsthorpe. Late on Christmas Eve in 1642, she went into labor, and around 2:00 on Christmas morning she gave birth to a tiny boy. He was so sickly at the time of his birth, that according to one account, two women attending the birth were sent for medicine. Instead of hurrying, they rested on the way back, assuming that the child would already be dead by the time they returned. In spite of this dire prediction, the child, who his mother described as being so tiny that he could fit into a quart pot, lived. She named him Isaac, after his late father.

Not much is recorded about the young Isaac's first years, but when he was three, his mother remarried. Her new husband, Barnabas Smith, was a well-off rector in nearby North Witham. He was substantially older than Isaac's mother, and there is reason to believe that the entire affair was carried out more as a business arrangement, than a romantic relationship. Smith's proposal was delivered, not in person, but by one of his servants.

The arrangements for the marriage were negotiated by the two families and included an agreement that Hannah's son would receive a parcel of land valued at fifty pounds, which he would inherit when he turned twenty-one. His mother would live with her new husband in North Witham, while Isaac remained with his grandparents in Woolsthorpe.

The effective loss of his mother at such an early age seems to have made a profound impression on Isaac. He was left in the care of his elderly grandparents and reportedly became an angry and withdrawn child. In a journal entry written when he was a teenager, Newton recorded a list of his sins. Under number thirteen he recorded, "Threatening my father and mother Smith to burn them and the house over them." The following entry contained the ominous, "Wishing death and hoping it for some" (Westfall).

Barnabas Smith died of natural causes in 1653, when Isaac was eleven years old. This facilitated the return of Isaac's mother, but any joy he might have felt at this reunion was tempered by the unwanted addition of two stepsisters and a stepbrother to his family. During the brief marriage, Smith had managed to sire three children, Mary, Hannah, and Benjamin. There seems to have been no love lost between Isaac and his new found stepsiblings. They quarreled frequently.

Having to compete for his mother's attentions was bad enough, but when he was twelve, it was decided that he should further his education by attending King's School at the market town of Grantham. Previously, he had gone to a local school, but Grantham was seven miles from his home, too far to walk daily. His family made arrangements for him to board with a family friend, the local apothecary, William Clarke. As much as the young Isaac might have objected to this second separation from his mother, it did provide him with opportunities that would shape his later life.

It was at King's School that Newton learned Latin and Greek, the languages of scholars. He also learned Hebrew as part of his extensive Bible studies. His education in these matters was helped along by his inheritance of Barnabas Smith's personal library. Smith was a pastor, and his collection contained approximately two hundred books. Newton gained access to these as well as to a largely blank notebook. Although he referred to it as his "waste book," Isaac would make special use of this notebook as his personal and scientific journal later. Along with grammar and literature, which formed the foundation of the King's School curriculum, it is also probable that while there, Isaac was introduced to the basics of mathematics.

In addition to his formal education, Newton spent many hours observing and assisting Clarke in his duties as the town apothecary. These consisted of preparing and measuring remedies of various sorts. Newton took this quite seriously and made extensive notes in his new notebook from books on the subject loaned to him by Clarke. Among the recipes that Isaac faithfully copied down was the following treatment for abscesses: "Drink twice or thrice a day a small portion of mint and wormwood and

300 millipedes well beaten (with their heads pulled off) suspended in 4 gallons of ale in fermentation" (Christianson, 2005).

When he wasn't by Clarke's side, rather than playing with local children, Isaac busied himself constructing models and devices of various sorts. He indulged his fascination with time and the movement of the sun by constructing sundials. These were reportedly so accurate that according to one of his biographers, "Anybody knew what o'clock it was by Isaac's dial, as they ordinarily called it" (Ackroyd, 2006).

After visiting the construction site of a windmill nearby, he built his own working model. This included a live mouse he called "the miller" who would trigger the wheels of the mill to begin turning when he reached for his food. He also developed a method of making paper lanterns and attaching them to kites. He was fond of lighting the small candles and flying the kites at night to alarm the villagers, who mistook them for comets.

Unfortunately, his interest and talent with mechanics did not translate into success at school. He described himself later as having "continued very negligent" in his studies. As a result, he was ranked second to last out of the school's eighty students. According to the story, this all changed one day as the result of a fight.

Isaac was small for his age, and one day one of the larger boys, of higher academic rank, kicked him in the stomach. Later that day, Isaac challenged the boy to a rematch after school in the nearby churchyard. Although his opponent was substantially larger than him, Isaac fought with such ferocity and determination that he beat the boy. In addition to sealing his victory by rubbing the defeated boy's nose against the church wall, Isaac also seems to have been galvanized by the confrontation to beat him academically as well. Newton began applying himself to his studies with similar ferocity and determination and soon became the school's top student.

All seemed to be going well for Isaac, but once again, matters were taken out of his hands. When he was fifteen, his mother decided that it was time for him to stop wasting his time with a useless education and begin taking on the responsibilities of a respectable landowner. To put it mildly, Isaac was disinclined toward the prospect. Despite his objections, and those of his school master, his mother insisted, and he returned once more to his family's estate in Woolsthorpe. However, if his mother had any illusions that Isaac would acquiesce and follow in his late father's footsteps, he did his best to quash them.

Instead of tending to his duties, young Newton would frequently wander off to sit under a tree, book in hand. This eventually led to his being fined in the local manor court for "suffering his swine to trespass in the corn fields" and "for suffering his sheep to break the stubbs" (Ackroyd, 2006) of other's fields. His mother responded by assigning one of the servants to supervise him, but Newton continued to shirk his responsibilities and returned to his reading or spent his time building more models.

It didn't take Newton long to gain the reputation for absent-mindedness, so frequently associated with genius. On one occasion, he was so lost in thought while leading his horse home, that when the horse slipped its bridle, he failed to notice. He returned home, bridle in hand, but without the wayward mount.

The battle of wills between Newton and his mother might have continued unabated without the intervention of his former school master, John Stokes, and Isaac's maternal uncle, William Ayscough. The two men tried to convince Hannah that trying to force Isaac into the role of rural landowner would be a waste of his talent and her time. To seal their case, Stokes offered to waive the forty-shilling tuition if Isaac returned to school. Reluctantly, Isaac's mother agreed. Newton returned to school, boarding with Stokes. He threw himself once more into his studies, and his academic success continued. He did so well, in fact, that when he graduated, Stokes delivered a speech to the student body, in which he tearfully praised the boy and urged the other students to follow his example.

Newton went on to Cambridge University in the summer of 1661, but while his mother may have resigned herself to his scholarly ambitions, she was certainly not going to pay a shilling more than necessary for them. Newton entered the august institution as a subsizar, a student who works his way through college acting essentially as a servant for the wealthier boys. His duties included waiting tables, running errands, and waking the other students before dawn so they could attend morning chapel.

Cambridge, like other universities of the time, based its curriculum on the classical philosophers. The seventeenth-century scholar Thomas Fuller described it as, "the stateliest and most uniform college in Christendom" (Ackroyd, 2006). There, Newton studied the works of Aristotle and Plato, including compulsory lessons in ethics, logic, and rhetoric. Newton quickly became bored with these. Like Paracelsus before him, Newton was chafing under the yoke of this hidebound education. He wrote in his notebook in Latin, "*Amicus Plato, amicus Aristoteles magis amica veritas*," which translates as, "I am a friend of Plato, I am a friend of Aristotle, but truth is my greater friend."

During this period, Newton continued his solitary tendencies.

He kept to his room and rarely associated with his fellow students. One exception was a classmate named John Wickins. The two met when Wickins was on a walk, apparently to get away from his roommates. He came upon Newton sitting on a bench, and somewhat uncharacteristically for both of them, they struck up a conversation. After commiserating about their fellow students, the two agreed to move in with each other. They remained together, Wickins often acting as Newton's secretary and assistant, for the next twenty years. There has been much speculation about the nature of their relationship, whether it was purely platonic or whether it might have been sexual. Unfortunately, neither Newton nor Wickins left any writings about their time together that would provide any clues.

As an undergraduate, Newton supplemented the university's prescribed reading with the works of more contemporary writers. These included the Polish astronomer Nicolaus Copernicus, whose works challenged the geocentric universe of Aristotle, and the German Johannes Kepler, who developed laws of planetary motion to explain the new heliocentric view. Newton also read the works of Galileo, who coincidentally died around the time of Newton's birth. The Italian's emphasis on mechanics, mathematics, and experimentation as tools for scientific exploration fascinated the young man. He took copious notes on all of these new ideas in his notebook.

It was also around this time that Newton began learning about higher mathematics by reading Descartes' *Geometry*. Like many fictional and historical geniuses, Newton was by nature autodidactic, a self-learner. According to one of his biographers, Abraham de Moivre, Newton:

Being at Sturbridge fair bought a book of astrology to see what was in it. Read it till he came to a figure of the heavens which he could not understand for want of being acquainted with Trigonometry. . . . Got Euclid to fit himself for understanding the ground of Trigonometry. Read only the titles of the propositions, which he found so easy to understand that he wondered anybody would amuse themselves to write any demonstrations of them.

He went on to master Descartes, and even went so far as to write in the margins, "Error—error non est Geom," but his slighting of Euclid would cost him later.

In the spring of 1664, Newton was required to pass an exam to receive a scholarship and qualify to sit for his bachelors of arts degree the following year. He was assigned to take the exam before Isaac Barrow, holder of the university's first Lucasian professorship in mathematics. It might be expected that this would be a snap for the young prodigy, but unfortunately, Barrow selected Euclid as the subject of the exam. Newton did manage to pass, but his performance was less than spectacular. The results of his examination the following spring were similarly unimpressive. In the vernacular of students of the time, he lost his groats. Traditionally, before the exam, students would give nine coins, or groats, to the official overseeing the exam. If the student did well, the coins were returned. If not, as in Newton's case, the coins were forfeit. In spite of this, Barrow must have seen some promise in the young scholar. He went on to become the young man's mentor and collaborator of many years. Newton received his scholarship and his BA. From then on, his place at Trinity was ensured.

Shortly after he received his degree, however, he was forced to leave Cambridge. In 1665, London fell victim to the plague. It quickly spread through the city's overcrowded neighborhoods and slums. By the end of January, it was killing hundreds each week. Within a short time, the weekly death toll was in the thousands. Anyone with the means soon fled London. Those unwilling or unable to leave were decimated. According to some estimates, before it had run its course, the plague killed one of every six Londoners. In what had once been a bustling metropolis, grass began to grow in the abandoned streets.

Officials in nearby Cambridge took note. The university was closed, and the faculty and students quickly retreated to the safer countryside. Newton returned to Woolsthorpe, with his books and there once more took up his independent studies. For some, this involuntary exile might have been a setback, but for a naturally reclusive genius like Newton, it simply provided him an opportunity to be alone with his thoughts, and what thoughts they were. As he would later put it, truth is "the offspring of silence and meditation."

While at Woolsthorpe, he continued his exploration of mathematics, finally mastering Euclid and once more taking up and soon surpassing Descartes. He soon filled the "waste book" he had inherited from his stepfather with seemingly endless calculations. At one point, he calculated the area under a parabola to fifty-five decimal places. To aid himself in his calculations of infinite and infinitesimal quantities, he created a whole new type of mathematics, what he called "the arithmetic of fluxions." Today we refer to it as calculus.

While still an undergraduate, Newton had begun a series of experiments into optics and the nature of light. These were likely inspired by his reading of Kepler's *Opticks*, and Robert Boyle's recently published *Experiments & Considerations Touching Colours*. He also read the newly published work of Boyle's former assistant, Robert Hooke. In the preface to his book *Micrographia*, Hooke said, "The Science of Nature has been already too long made only a work of the Brain and the Fancy. It is now high time that it should return to the plainness and soundness of Observations on material and obvious things."

Regardless of what inspired them, Newton continued his investigations in Woolsthorpe and threw himself into them with a passion. To discover the effects of light he would stare at the sun for extended periods of time and then make extensive notes on the colored spots and circles that appeared afterward when he looked at a dark wall. While this gave him valuable insights into the nature of color, it also nearly succeeded in blinding him. He soon found that the spots persisted long after he had stopped staring at the sun. He was forced to shut himself in a dark room for several days until his vision recovered.

One would think that this would have caused him to exercise more caution in his research, but the opposite seems to have been true. To determine if colors where being produced by the eye itself, he slid a blunt needle, or bodkin, between his own eye and the bone of his eye socket. He found that when he used the needle to apply pressure to the eye, it caused several white, dark and colored circles to appear to him. When he ceased applying the pressure, the circles disappeared.

Fortunately for science, and Newton's continued vision, he purchased a glass prism from the nearby Stourbridge Fair the year before, and he focused his attentions on that as a means of studying light. He drilled a small hole in the shutters of his room and directed the beam of sunlight that entered to pass through the prism. Instead of the perfectly circular rainbow pattern predicted by Descartes, Newton saw on the opposite wall the familiar colors of red, orange, yellow, green, blue, indigo, and violet arranged in an oblong. This was not due, as most at the time believed, to the prism creating the colors, but as Newton discovered, it was due to the way that the prism bent the incoming white light. It did so at specific angles, and Newton measured these and used his emerging mathematical skills to calculate the angle that corresponded to each color.

At this point, Newton took his discovery one step further, and performed what he described as his *experimentum crucis*, crucial test. He directed the individual colors produced by the prism through a second prism. Instead of splitting into further colors, they passed through the new prism unchanged. The red beam remained red. The blue remained blue, because they were, as Newton described them, pure colors. The white light that passed through the original prism, on the other hand, was a heterogeneous mixture, which the prism merely separated into its pure components.

When not studying light in his darkened room, Newton spent time in the garden of his Woolsthorpe home. The garden contained an apple tree, and while the story of Newton's being inspired by the falling of an apple is likely apocryphal, he did find time to ponder gravity. The falling of objects had long been observed and studied. Galileo had made extensive studies on the subject, which Newton had read. Newton's intuitive leap was in realizing that the force that caused things like apples to fall was the same force that kept the planets in their paths around the stars. This insight didn't come in the instantaneous flash that schoolbooks frequently report, but Newton did lay the ground-work at Woolsthorpe for what would eventually become his law of universal gravitation.

During this period, which many historians have dubbed Newton's *annus mirabilis,* or wondrous year, he added to his reputation as a mad genius, not only by the depth and breadth of his discoveries, but also by the single-minded dedication with which he pursued them. One of his disciples, Dr. George Cheyne, later reported that while he was pursuing his research on optics, "to quicken his faculties and fix his attention, he confined himself to a small quantity of bread, during all the time, with a little sack and water, of which, without any regulation, he took as he found a craving or failure of spirits."

Newton returned to Cambridge in the spring of 1667, but rather than trumpeting his discoveries upon his return, he kept them largely to himself. Perhaps he lacked confidence. He was only twenty-four years old at the time. Perhaps he liked the idea of holding secrets that he alone possessed. For whatever reason, he guarded his newfound knowledge jealously and revealed it to few. One of those with whom he shared his mathematical breakthroughs was his mentor Isaac Barrow. Being a mathematician, the Lucasian professor recognized the significance of them and urged his young protégé to publish. Newton refused but Barrow persisted.

Two years later, Newton was still resisting, but Barrow received from one of his colleagues in the Royal Society, John Collins, a copy of the book *Logarithmotechnia* by the German, Nicolaus Mercator. It laid out a simplified method for calculating logarithms, but Barrow realized that this was what Newton had done three years earlier. The professor used this to pressure Newton to record his work at last, and he reluctantly agreed. He wrote an account titled *On Analysis by Infinite Series*, but refused to have it published. He allowed Barrow to send a copy to Collins, but only if the work remained anonymous. Barrow wrote to Collins and told him, "A friend of mine here, that hath a very excellent genius to these things, brought me the other day some paper, wherein he has set down methods of calculating the dimensions of magnitudes like that of Mr. Mercator concerning the hyperbola, but very general" (Christianson, 2005).

When Collins received this work, he was greatly impressed and responded enthusiastically. It was only then that Newton allowed Barrow to reveal his name. The young genius was still resistant to publishing, but Barrow, now with Collins on his side, was making progress. In addition, they began discreetly circulating word of Newton's breakthroughs among the mathematical community. Intrigued, the mathematicians began sending inquiries and problems to Newton via Collins. His insightful replies only added to his budding reputation.

Meanwhile, Newton happily ensconced again in his beloved Cambridge, finished his masters degree. Barrow, who in addition to being a mathematician was a prominent theologian, was offered the position of chaplain to King Charles II. He gladly accepted the position and used his influence to ensure that Newton filled his seat as Lucasian professor. The young Newton now had a tenured position, which could be taken away only for the crimes of fornication, heresy, or voluntary manslaughter. It provided a stipend, of one hundred pounds per year, which together with the rents from the property he had inherited, left him financially quite comfortable. Add to that the acclaim of his mathematical peers, and the young man's future was now secured. He was twenty-seven years old.

Newton made his first trip to London late in 1669, but not to see the sights. Most of them had been destroyed in the great London fire three years earlier. No, he traveled with a more revealing purpose in mind. He made his way to Gray's Inn Lane and the Swan tavern. While there, he began making contacts with the city's informal network of alchemists. Soon he was frequenting a small bookstore at the sign of the Pelican in the section of the city known as Little Britain. There he made the acquaintance of William Cooper, a dealer in legal and less-than-legal manuscripts. Through him, Newton procured numerous tracts on the ancient art, including the six-volume *Theatrum Chemicum* by Lazarus Zetzner and *Ripley Reviv'd* by Eirenaeus Philalethes.

While in London, Newton also purchased numerous alchemical supplies, including aqua fortis, sublimate, oyle perle, fine silver, and antimony, among others. He bought two furnaces, one made of tin and the other of iron, and had these shipped back to Cambridge. Upon his return, he quickly set up a laboratory in the quarters he shared with Wickins, and earnestly undertook alchemical experiments. As some small indication of the dedication with which he took to this work, before the age of thirty, Newton's hair had turned prematurely gray. When Wickins asked him if this was the "effect of his deep attention of mind," Newton half seriously told his friend that he had so frequently experimented with quicksilver, the alchemists' name for mercury, "as from thence he took the Colour."

Eventually, Newton's experiments became elaborate enough that he was forced to hire an assistant. He chose for this position a distant relative from Grantham, Humphrey Newton. The young Humphrey painted a detailed, if somewhat chilling, picture of his master's pursuit of these experiments:

> So intent, so serious upon his studies that he eat very sparingly, nay ofttimes he has forgot to eat at all. . . . He very rarely went to Bed till 2 or 3 of the Clock, sometimes not till 5 or 6, especially at spring & fall of the Leaf, at which Times he used to imploy about 6 weeks in his Elaboratory, the fire scarcely going out either Night or Day, he sitting up one Night, as I did another, till he had finished his Chymicall Experiments, in ye Performance of which he was the most accurate, strict, exact.

Later, Humphrey continued, "But what his Aim might be, I was not able to penetrate into, but his Pains, his Diligence at those set Times made me think he aimed at something beyond the Reach of human Art & Industry" (Christianson, 2005).

It would be difficult to more accurately capture the popular image of a mad scientist, but we must be careful not to imagine this as all being due to some sort of schizophrenic split in Newton's personality, the rational scientist on one side and the crazed alchemist on the other. What most people think of as the science of chemistry was still in its embryonic stages. It had been conceived by men like Robert Boyle, but its actual birth would have to wait until the following century, when others, like Joseph Priestley and Antoine Lavoisier would help it into the world. In the 1600s, those who wished to unlock the secrets of matter, like Newton and Boyle before him, would need to master the work of the alchemists. In fact, the young Isaac Newton would be helped along the way by Robert Boyle himself.

Newton and Boyle met publicly for the first time when Newton visited the Royal Society in 1675. Many have speculated that the two may have met informally through the private community of alchemists that both men were members of; however, there is little if any documentation of this. What is known is that Boyle, who at the time was one of the most prominent scientists in England, warmly met the quickly rising Newton. They began corresponding regularly shortly after that. Newton showed particular interest in Boyle's thinly disguised report in the *Philosophical Transactions of the Royal Society* on his production of philosophical mercury. Newton immediately recognized the alchemical significance of this work and warned the senior scientist of the dangers of making such knowledge available to the public. Whether Boyle listened to his new friend's advice is unknown, but the two remained relatively close for many years to come.

The relationship between Newton and Boyle's former assistant Robert Hooke was, however, substantially less cordial. In fact the intellectual war that ensued between the two was worthy of any fictional mad scientist and the battles with his arch nemesis. Hooke was by this point a prominent scientist in his own right, and held the position of curator of experiments within the Royal Society. This suited his talents well because, while he may have lacked the mathematical mastery of Newton, he was a brilliant experimentalist who made important discoveries in a wide range of fields, including anatomy, microscopy, chemistry, and meteorology.

Although both Newton and Hooke were quite accomplished, both dedicated to uncovering the workings of the universe, it is hard to imagine two more different individuals. Newton was an almost pathological loner, who obsessively pursued his research with single-minded intensity, denying himself any creature comforts that might distract him from his work. Hooke, on the other hand, seemed almost to flit casually from one area of research to another. He was also gregarious by nature, reveling in the social life and gossip of the coffeehouses of the time, and rarely denying himself the pleasures of fine food or drink. In addition, while Newton adhered to his Puritanical beliefs, Hooke, although never married, was known to have a succession of mistresses, and would frequently write about them in his journal, along with descriptions of the quantity and quality of his orgasms. The instantaneous dislike between the two so disparate men was almost instinctual.

They first came to intellectual blows when Newton's work on optics was presented to the Royal Society. As curator of experiments, it was expected that Hooke would review it before it was formally presented. His initial report was less than complimentary, almost dismissive of Newton's work. This may have been due to the cursory reading he gave the paper. He later admitted that he had allowed himself only about three hours to review what by most standards was a rather complex work. It may also have been due to his inability to fully grasp all of Newton's complex mathematics. It should also be said that Newton's findings contradicted several of Hooke's own beliefs about light which he had laid out in his own work *Micrographia*.

For Newton, who was extremely resistant to publishing to begin with, this negative reaction from a prominent man of science may have been his worst nightmare come true. He was back in Cambridge at the time, and responded with a letter, that while cautious and polite in tone, was unmistakably angry. Hooke countered with further attacks, claiming that Newton had merely posed an interesting hypothesis without providing sufficient proof. The battle was under way, with fellow scientists weighing in on both sides.

Eventually, however, Newton, with the support of some very influential Royal Society members, was able to refute each one of Hooke's accusations, and the society determined that Hooke should reconsider the original work. In his response, Newton made his famous statement, "If I have seen farther, it is by standing on the shoulders of giants." This sounds magnanimous, but it may have actually been a not too subtle jab at Hooke who, due to a curvature of the spine, was actually quite short.

The battle may have been over, but it would prove to be only the first skirmish in a protracted war of science. Hostilities flared again when Newton was working on his masterpiece, perhaps the most influential scientific work of all time, *Philosophiae Naturalis Principia Mathematica*. It would come to be known around the world by the simpler name *Principia*. The scientist Edmund Halley had started wondering about the effects of gravity. Specifically, he wondered what the effect on an orbiting body would be if the force of gravity on it adhered to what was called the inverse square law. What if the force of gravity were the inverse of the square of the distance between the orbiting body and the body it orbited around? In other words, if the distance between them were doubled, multiplied by two, would the force of the pull between them be one over two squared, or one quarter, as strong, and what would that do to the shape of the orbit?

It may seem to be an esoteric sort of question, but it has important implications for the movement of planets, and no one that Halley posed the question to, including Robert Hooke, could come up with sufficient mathematical proof to answer it.

Halley decided to journey to Cambridge and ask the question of Newton. To his complete amazement, Newton immediately came up with an answer. He responded that this would cause the orbit to be elliptical rather than round. Dumbfounded, Halley asked him how he knew, and Newton said that he had already calculated it, and began rooting around in his study for the notes. Unfortunately, he was unable to lay his hands on them, but he assured Halley that he would redo the calculations and send them to him immediately.

Three months later, Halley was still waiting. It wasn't that Newton hadn't done the calculations. It was that, when he went to redo them using his new mathematical techniques, he had become intrigued with the problem and thrown himself once more into it with characteristic obsession. He wrestled with it almost nonstop until 1684 when he sent Halley a nine-page manuscript he called *De Motu Corporum in Gyrum* (On the Motion of Revolving Bodies).

When Halley opened the package and read it over, he realized that he held in his hands a work of genius. Here were the mathematical seeds of an entirely new branch of science dealing with motion and the forces that affect it, what today we call dynamics. He immediately told the Royal Society of this breakthrough and urged Newton to publish as soon as possible. Unfortunately, once Newton was focused on a problem, he did not release it that easily. He wrote back to Halley to say, "Now that I am upon this subject, I would gladly know the bottom of it before I publish my papers" (Christianson, 2005). It would take Newton eighteen months to reach the bottom to his own satisfaction, but when he was done, he presented to the Royal Society the first third of what would become his great work, *Principia*.

Almost predictably, Hooke reacted to his rival's great triumph with charges of theft. He claimed that he had developed the inverse square law himself six years earlier and that Newton was trying to take credit for his work. When word of this reached Cambridge, Newton was justifiably outraged. He wrote to Halley and, in his anger, threatened to withhold the rest of the work. Fortunately for the history of science, Halley was able to mollify the young Newton. Hooke could make his claims, but he didn't have the mathematical proof necessary to back them up. With that assurance, Newton completed the work. In the first of its books, he dealt with motion in the absence of resistance or friction. In the second, he covered objects moving in fluids. In the third and most influential of the books, he laid out his famous three laws of motion.

That was far from the last time that Newton and Hooke would cross swords. Their bitter rivalry continued until Hooke's death in 1703. There would be others, of course, who would challenge Newton, among them the royal astronomer John Flamsteed and the great German mathematician Gottfried Wilhelm von Leibniz. In the end, however, Newton would go on to become president of the Royal Society, as well as a Member of Parliament and master of the English mint. He is widely acknowledged as one of the greatest men science has ever known.

Newton died in 1727, at the venerable age of eighty-four. He was laid to rest in Westminster Abbey with Britain's highest honors. In 1936, the Viscount Lymington, a descendent of Newton, auctioned off a metal chest containing a large volume of Newton's writings. It was purchased by the economist John Maynard Keynes. When he began looking through the pages, what he found there amazed him. They contained the story, not of the cold rational Newton so often portrayed in textbooks, but of a passionate genius, an alchemist, an isolated man, habitually on the cutting edge. As Keynes later told his students at Trinity College, "Newton was not the first of the age of reason, he was the last of the magicians, the last of the Babylonians and Sumerians, the last great mind which looked out on the visible and intellectual world with the same eyes as those who began to build our intellectual inheritance rather less than 10,000 years ago."

CHAPTER THREE

Transmutation

From Alchemy to Chemistry

THE ENLIGHTENMENT, THE PERIOD IN WHICH REASON
seemed to make unprecedented leaps and bounds in every area of
human enterprise, was at its height by the middle of the eighteenth
century. It had been helped along the way, in no small part, by
Newton's *Principia*, as well as the writings of men like Descartes
and Locke. Philosophy, economics, politics, religion, and especially science were all seen as if with new eyes, eyes willing to re-
examine old assumptions and make new insights. The results were
revolutionary.

It might seem that the victory of reason over superstition
would spell the end of our fear of science and those who wielded
its tools. On the contrary, much of the mystique of the alchemists
was spilling over into the bubbling glassware of the chemists. As
they began applying their new investigatory tools to dispel the
trusted beliefs of the ancients, many considered it a direct threat

to all they held dear. Every new discovery was seen as a new volley in the war of tradition versus heresy.

Onto this battlefield stepped an unlikely figure. He was a humble clergyman with a slight stutter from a small parish in rural England, but he would go on to make his own revolutionary contributions in science as well as in religion, linguistics, education, and politics. He would also go on to influence some of the leading thinkers of the day on both sides of the Atlantic and come to be seen as such a threat to his native England that he would, in the tradition of many a mad scientist, have to flee for his life from an angry mob. His name was Joseph Priestley.

Joseph Priestley was born in March of 1733, only five years after the death of Isaac Newton, in the small town of Fieldhead near Leeds. He was the oldest child of wool merchant and nonconformist Christian preacher, Jonas Priestley. At the time, the term *nonconformist Christian* included Calvinists, Quakers, Presbyterians, and anyone else not part of the Church of England. Joseph would go on to follow in his father's dissident religious footsteps.

His mother died when he was nine, and his father, being unable to properly care for the boy, sent him off to be adopted by his childless aunt, Sarah Keighley. She and her husband were sympathetic to the dissenters' cause and would frequently play host to their meetings. Around their dinner table, Joseph was exposed to many a lively discussion of theology and politics.

They also saw to it that Joseph had a first rate education. He began studying for the clergy when he was twelve, showing himself to have a fine intellect and a particular talent for languages. He learned Latin, Greek, and Hebrew as part of his formal studies, and during a period when he contracted tuberculosis and was forced to leave school, he taught himself French, German, and Italian. He also followed Newton's example by schooling himself in the basics of mathematics, geometry, and algebra.

Priestley recovered and finished his studies. By the time he was nineteen, he was ready for college. Unfortunately, Oxford and Cambridge at that time were reserved exclusively for members of the Church of England, so he attended Daventry Academy, the best of what were known as the Dissenting colleges.

Priestley avidly applied himself at college and graduated in three years, rather than the traditional four. Upon graduation, he accepted a position as pastor of a small congregation in Needham Market. Unfortunately for Priestley, his religious views, particularly his questioning the divinity of Jesus Christ, proved too radical even for the Dissenters. These views would eventually lead Priestley to become one of the founders and leading spokesmen for the Unitarian Church, but as young preacher in a rural parish, they were a bit much. He moved on after only three years.

Eventually, the young preacher found a position with a small congregation in Nantwich that was a better fit. It consisted of only sixty parishioners, leaving Priestley plenty of time for other pursuits. In addition to his regular duties, he was put in charge of the parish school. He took to this readily, and was soon teaching the school's thirty boys and six girls the basics of reading, writing, and arithmetic.

He also found the time to do some writing. Being disappointed with the grammar texts available to his students, he wrote his own. It was titled *The Rudiments of English Grammar* and was one of the fist attempts to examine the English language using the same rigor and scholarship long applied to Latin and Greek. This unconventional new approach proved quite successful and encouraged him to continue writing, notably on the subject of education. Priestley advocated adding subjects such as modern history, economics, and political science to the standard classical curriculum and was an early and vocal proponent of public education. He found a ready audience for these views and was awarded a doctorate of law degree from the University of Edinburgh for his contributions to education.

With his newfound success, Priestley was able to land a more lucrative position as professor of languages at the Warrington Academy in Yorkshire. Aside from his regular classes, he began tutoring the young William Wilkinson, son of the wealthy industrialist John Wilkinson. Sometime during the course of tutoring William, Priestley became acquainted with the boy's sister, Mary. He found himself taken with the intelligent and personable young woman, and the two were soon married.

Before becoming a teacher, Priestley had little in the way of formal science education, but he viewed science as an important part of his students' education. As he wrote in one of his later works:

> I am sorry to have occasion to observe, that natural science is very little, if at all, the object of education in this country, in which many individuals have distinguished themselves so much by their application to it. And I would observe that, if we wish to lay a good foundation for philosophical taste, and philosophical pursuits, persons should be accustomed to the sight of experiments, and processes, in early life. They should

more especially be early initiated in the theory and practice of investigation, by which many of the old discoveries may be made to be really their own.

With this in mind, he began educating himself in the topic to better instruct his young charges. He even purchased a few scientific instruments for their benefit, including an air pump, a well-calibrated scale, and a machine for generating static electricity.

The scientists of the time were a far cry from the professional researchers of today. In fact, the word *scientist* wouldn't come into popular English usage for another hundred years. Instead, scientific experiments in the 1700s and well into the 1800s were carried out largely by amateurs, dilettantes, and dabblers who had the time and inclination to explore the scientific mysteries of the day. Once he began conducting and demonstrating experiments for his students, Priestley quickly became hooked and was soon on his way toward becoming one of the dabblers.

Like many of his peers, Priestley became fascinated with the newly fashionable phenomenon of electricity. It was quite in vogue at the time, due in no small part to the efforts of an American experimenter named Benjamin Franklin. He and his fellow electricians, as they were called, were busy capturing the public's imagination with flamboyant demonstrations of electrical wonders. What's more, they were inventing practical applications for their electrical knowledge, including Franklin's lightning rod and E. G. von Kleist's invention of an early capacitor known as a Leyden jar.

Priestley was struck by the fact that no one had documented the history of this rapidly expanding field. He was still enjoying the success of his earlier book and thought he might be just the man to tackle the job. With that goal, he set off for London to meet the electricians for himself and propose to them that he chronicle their discoveries. To smooth the way, he brought with him a letter of introduction from the rector of the Warrington Academy, John Seddon. It was addressed to one of England's leading electricians, John Canton, a member of the Royal Society. In his letter, Seddon wrote, "You will find [Priestley] a benevolent, sensible man, with a considerable share of Learning." At the end, there was a short postscript: "If Dr. Franklin be in Town, I believe Dr. Priestley would be glad to be made known to him." (Johnson)

As it happened, Franklin *was* in town. He was serving as the deputy postmaster general for North America. During the extended stays in London this required, Franklin would often frequent one of the local coffeehouses near St. Paul's Cathedral, where he could strike up conversations with fellow freethinkers. Eventually, Franklin and his friends formed an informal group they called the Club of Honest Whigs. They would meet on alternate Thursdays and discuss the burning issues of the day over coffee and porter. Frequently these discussions would involve scientific questions, and new theories and speculations about electricity were a frequent topic.

It was just such a meeting that John Canton attended in 1765, and invited the young Dr. Priestley to tag along. Soon the humble pastor was sitting across the table from his heroes, including the world-famous Dr. Franklin himself. Franklin and the others welcomed their new admirer warmly as one of their own, and over the course of the afternoon, Priestley laid out his book proposal. Not only did they give the idea a warm reception but promised to help by supplying research materials and offering to read the manuscript. They also suggested experiments that he could conduct himself to better understand the subject. With their encouragement, Priestley leaped fully into the world of science.

Within a few days, Priestley was accompanying Franklin and Canton as they visited the Royal Society. Imagine what it must have been like for Priestley to enter those hallowed halls of science where Newton himself had once presided, and in such august company. In less than a week, he had gone from a young dabbler to an honored guest of the great men of science. He would soon be sitting among them as a scientific peer.

When his time in London was done, he rushed back to Warrington and began assembling a sort of makeshift laboratory, using his limited funds to purchase the tools and materials he needed. He was soon throwing himself into scientific experimentation with an almost manic intensity. Unfortunately, there is no record of his wife's reaction to all this, but she must have been less than pleased when, to conduct some of his messier experiments, he commandeered their kitchen sink.

Most of these early experiments were his attempts to reproduce the results of others, but then he started designing experiments to answer his own questions. To modern readers, the descriptions of these might sound like the script for a mad science movie. In one he described to his new scientific colleagues, Priestley used glass tubes coated with foil to deliver electric shocks. He wrote:

I have made a great number of experiments on animals, some of which I refer to in a letter I lately wrote to Dr. Watson. Since I wrote to him, I discharged 37 square feet of coated glass through the head and tail of a Cat three or four years old. She was instantly seized with universal convulsions, then lay as dead a few seconds. . . . Thinking she would probably die a lingering death in consequence of the stroke, I gave her a second, about half an hour after the first. She was seized as before, with universal convulsions, and in the convulsive respiration which succeeded she expired. She was dissected with great care, but nothing particular was observed (Schofield, 1966).

Science in the eighteenth century was not constrained by the niceties of animal welfare laws, but at the time, experiments such as these served an important purpose laying the groundwork for future discoveries. This and other experiments lead Priestley to propose that there was an inverse square relationship for electrical charges, similar to Newton's inverse square law for gravity. In other words, as the distance between two charged objects decreased, the strength of the charge between them increased exponentially. A few decades later, the French physicist Charles-Augustin de Coulomb conclusively proved this was true. Because he provided proof of Priestley's supposition, it would become known as Coulomb's law.

Although Priestley missed having a scientific law named after him, by the summer of 1766, Franklin, Canton, and several other members of the Honest Whigs were sufficiently impressed by his efforts to nominate their young friend for membership in the Royal Society. Priestley's position within the scientific community was now ensured.

Incidentally, it should also be mentioned, that during this period Priestley made one of his least known, but longest lasting, contributions to science and society. He was working on the manuscript for his book when he discovered that a newly imported type of tree sap from South America was just the thing he needed to rub out errant pencil marks on the paper. He therefore called the new substance "rubber."

Only fifteen months after his first introduction to the Honest Whigs, Priestley completed the manuscript for his electrical history. It contained an exhaustive history of electrical discoveries, including a full description of Franklin's famous kite-in-a-lightning-storm experiment and advice for aspiring electricians. In the last two hundred pages, Priestley gave descriptions of his own electrical discoveries.

Published in 1767, Priestley's *The History and Present State of Electricity*, was warmly received. It sold well and was lauded in reviews as "excellent" and "well-informed." It went through five editions in England and was translated into French and German. Franklin made sure that copies were sent to the American colonies. It was soon considered one of the primary textbooks on electricity and would remain so for more than one hundred years.

Shortly after the publication of his great book, Priestley had a stroke of luck. This was the second in a series of lucky strokes that helped him become one of the great men of science. The first had been his meeting with Benjamin Franklin. Without that meeting, he might have continued as a dabbler on the outskirts of science. Franklin inspired, encouraged, and supported him and, in the process, shaped his political and scientific views. Priestley's second lucky break occurred in the summer of 1767, when he moved next to a brewery.

The move was necessitated by the growing demands of Priestley's family. He and Mary had a daughter, Sally, who was by then four years old, and even with his additional writing income, the salary at Warrington was insufficient. To better provide for his growing family, he accepted a position as minister to the Mill-

Hill Chapel in Leeds. This was a large, prosperous parish, and the job not only provided a decent salary but also a nice house on Bansinghall Street. As luck would have it, though, the house was still in the process of being renovated when it came time for the family to move. While the Priestleys waited, they rented a small house on nearby Meadow Lane next to the Jakes and Nell Public Brewery.

Ever curious, Priestley was soon poking around the brewery and talking to the workers. He discovered that while they were fermenting, the vats of beer produced vast quantities of a gas, called mephitic air, discovered only a few years earlier by the Scottish chemist, Joseph Black. It was what today we call carbon dioxide. This new gas was a hot topic of scientific discussion at the time, almost as much so as electricity, and Priestley was soon directing the majority of his research toward the new substance. The workmen must have wondered about the sanity of the eccentric scientist bent over their vats taking samples and conducting experiments.

In what Priestley described as one of his happiest discoveries, he found that by mixing water back and forth between two cups over the vats, he could make the water fizzy. This reminded him of certain expensive mineral waters popular at the time, and he described the process in a letter to his friend Canton, "I make most delightful *Pyrmont Water*, and can impregnate any water or wine &c. with that spirit in two minutes." Joseph Priestley, author, educator, theologian, and scientist could also be described as the father of carbonated soft drinks.

Given the multibillion dollar industry that soft drinks are today, one would think that Priestley's invention would have made him quite wealthy, but he believed strongly in a world in which scientific ideas were shared freely and proprietary secrets were unknown. He revealed his technique quite freely in a pamphlet titled "Directions for Impregnating Water with Fixed Air, in Order to Communicate to it the Peculiar Spirit and Virtues of Pyrmont Water, and Other Mineral Waters of Similar Nature." He thereby forfeited a fortune and most of the money to be made from soda at the time went to the man who invented the machine for carbonating water on a commercial scale, a fellow named Schweppe, of future gin and tonic fame.

Although he didn't make a lot of money from it, Priestley's early experiments with carbon dioxide did serve the purpose of getting him intensely interested in gasses. It was that area, known as pneumatic chemistry, in which he made some of his greatest discoveries. He went on to fully describe the properties of mephitic air (carbon dioxide) and discover no fewer than nine entirely new gases, including ammonia, hydrogen chloride, sulfur dioxide, silicon fluoride, and nitrous oxide.

When his family was eventually able to move into the pastor's house on Bansinghall Street, it cost him his easy access to the beer vats, but Priestley soon found other ways of producing mephitic air for his studies and constructed a pneumatic trough for his work. This device, originally invented half a century before by chemist Stephen Hales, consisted of a wooden box, two feet long and nine inches deep. At one end of the box there was a shelf with holes cut in it. The trough was filled with liquid to just above the level of the shelf. This liquid was usually water, but Priestley found that if the experiment involved water-soluble gasses, mercury could be used instead. Using this device, he could place inverted glass containers on the shelf, and the liquid would create a seal to trap the gases inside. He could then place things into the containers through the holes in the shelf.

While studying respiration, for example, Priestley would often use a mouse, which he grabbed by the scruff of the neck and deftly passed into the water, through one of the holes and into the container. He could then measure how long it took the trapped mouse to expire, again, so much for animal welfare. These experiments, however, weren't undertaken out of morbid curiosity. At the time, there wasn't a clear understanding of exactly what was happening. Did the animal exhaust the supply of "good" air in the container or did it somehow pollute the environment with "bad" air? In either case, the mouse always died. The air remaining in the container always appeared to be visually unchanged, but candles would refuse to burn in it. In fact, by placing a burning candle in the container with the hapless mouse, Priestley observed that the animal died faster, at which point, the candle promptly went out.

Priestley's breakthrough came when he tried a variation on the experiment in 1771, by using a mint plant instead of the mouse. Would the plant last longer than the mouse, or would it die faster? While the mice died within a matter of minutes, the plant in the jar survived for hours, which turned into days, and eventually into months. No matter how many times he repeated the experiment, or which type of plant he used the results were the same. Priestley described it:

> The plant was not affected any otherwise than was the neces-
> sary consequence of its continued situation; for plants grow-
> ing in several other kinds of air, were all affected in the very

same manner. Every succession of leaves was more dimin-
ished in size than the preceding. . . . The root decayed, and
the stalk also, beginning from the root; and yet the plant
continued to grow upward, drawing its nourishment through
a black and rotten stem.

What was more confusing, when he placed a candle into the
container which had contained the mint and lit it using a lens, it
burned fiercely, unlike the candle in the container with the mouse.
Eventually, the candle would go out, but unlike the mouse, the
mint continued to thrive. If he placed a mouse in with the mint,
the animal survived much longer than a mouse without the
mint. It was as if the plant were not only able to survive the bad
air but also able to transform the bad air into good air. Priestley
excitedly described his results in a letter to Franklin:

I have fully satisfied myself that air rendered in the highest
degree noxious by breathing is restored by sprigs of mint
growing in it. You will probably remember the flourishing
state in which you saw one of my plants. I put a mouse [in]
the air in which it was growing on the Saturday after you
went, which was seven days after it was put in, and it continued in it five minutes without shewing any sign of uneasiness,
and was taken out quite strong and vigorous, when a mouse
died after being not two seconds in a part of the same original
quantity of air, which had stood in the same exposure without a plant in it. The same mouse also that lived so well in the
restored air, was barely recoverable after being not more than
one second in the other. I have also had another instance of a

mouse living 14 minutes, without being at all hurt, in little more than two ounce measures of another quantity of noxious air in which a plant had grown.

A frenzied series of letters between Priestley and Franklin followed in which the Englishman used his American mentor and colleague as a sounding board to unravel the implications of this discovery. By 1774, Priestley was ready to publish his results in a book titled *Experiments and Observations on Different Kinds of Air*. What he laid out was a revolutionary theory:

> Once any quantity of air has been rendered noxious by animals breathing in it as long as they could, I do not know that any methods have been discovered of rendering it fit for breathing again. It is evident, however, that there must be some provision in nature for this purpose, as well as for that of rendering the air fit for sustaining flame; for without [it] the whole mass of the atmosphere would, in time, become unfit for the purpose of animal life; and yet there is no reason to think that it is, at present, at all less fit for respiration than it has ever been. I flatter myself, however, that I have hit upon one of the methods employed by nature for this great purpose. How many others there may be, I cannot tell.

Priestley had uncovered the vital symbiotic partnership between plants and animals. The idea that plants clean the atmosphere and provide us with breathable oxygen is now familiar, but in Priestley's time, it was groundbreaking, and laid the foundation for what would become the science of ecology. The Royal Society was rocked by the significance of Priestley's work and, in recognition, awarded him their highest honor, the Nobel Prize of its day, the Copley Medal.

This catapulted Priestley into the upper strata of scientific minds and attracted the attention of powerful admirers. One was the former secretary of state William Petty, Earl of Shelburne. He was so impressed with Priestley's scientific accomplishments and liberal politics, that he made him an extraordinary offer. In exchange for acting as Shelburne's adviser, organizing his library, and overseeing the education of Shelburne's two sons, Priestley was offered a home for himself and his family at the earl's estate in Calne, a sizable allowance of 250 pounds per year, and his own laboratory. It was an offer too good to refuse.

His new lab was a far cry from the one he had cobbled together in Mary's kitchen. It was housed in its own building and equipped with the finest scientific instruments, including a twelve inch convex lens or burning glass that had once belonged to Grand Duke Cosimo III of Tuscany. With tools such as these, Priestley was able to make perhaps his greatest discovery; however, it would lead to his greatest controversy and also one of his most famous mistakes.

Priestley immediately put his new tools to use. In August 1774, he turned the grand duke's lens on a reddish powder called mercury calx. Now called mercuric oxide, it's an ash created when mercury is heated in air. Priestley isolated a sample of the powder under a bell jar, then focused the lens on it. As the rays of the sun heated it, it produced a large quantity of an unknown gas. Priestley was unsure what the new gas was, but he suspected that it was the mephitic air with which he was so familiar. To test this, he put a lit candle into the jar, fully expecting it to go out. Instead if flared to life and burned with a never before seen intensity. He repeated the procedure with similar results. When he put a smoldering wooden chip into the new gas, it burst into flames. A heated wire would glow white hot in the mysterious substance. What was the identity of this newfound air?

Just as Priestley was trying to answer that question, duty called. He was asked to set aside his experiments temporarily and accompany Shelburne on a trip to Europe. Priestley and his benefactor toured the great cities. After landing in Calais, they completed a circuit, traveling through Belgium, the Netherlands, Germany, and then back to France. In September 1774, they reached Paris.

Although Priestley was wandering beyond the confines of England for the first time, he had been given an honorary membership in the French Academy of Sciences two years earlier for his contributions to science. Once in Paris, the French intelligentsia welcomed him with open arms. He was wined and dined in lavish fashion, and it was at one of these dinners that he was introduced to an ambitious young French chemist named Antoine Lavoisier.

The relationship between England and France at this time was at best strained. The two countries had a long history of hostility toward one another, and even when they were not actively at war, both maintained networks of spies to try to capture the secrets of the other. Into this atmosphere of intrigue and treachery, Priestley stepped with characteristic, if completely naive, openness. Over the course of dinner, he casually described all of his most recent discoveries, including the new gas. In recounting the conversation, Priestley said, "I never make the least secret of any thing that I observe." The French scientists, particularly Lavoisier, listened with rapt attention.

Although enjoying the attention, Priestley eventually tired of France, and returned home. Once back in his laboratory, he again set about learning more about his mystery gas. He tried various methods and produced even purer samples, and it occurred to him that if this new air could support combustion, perhaps it could support respiration as well. To test this, he placed a mouse into a container of the gas. Remarkably, rather than dying after a few seconds, as the mice typically did in mephitic air, this mouse remained alive and quite comfortable even after half an hour. He repeated the procedure and the results were the same.

Intrigued, Priestley performed the standard experiment for any fictional mad scientist. He tried the new substance on himself. He described the results in his journal:

> The feeling of it to my lungs was not sensibly different from that of common air; but I fancied that my breast felt peculiarly light and easy for some time afterwards. Who can tell but that, in time, this pure air may become a fashionable article in luxury. Hitherto only two mice and myself had had the privilege of breathing it.

He had not only discovered oxygen but had postulated the future existence of O_2 bars and had lived to tell the tale.

Priestley formally notified the Royal Society of his new discovery in March 1775. He described the properties of his new air, but he did not call it oxygen. That name would come later, at the hands of another. Priestley used the terminology of the time, and came up with a name for the new substance based on its ability to make things burn.

The accepted wisdom of the day was that things burned because they contained a mysterious substance called phlogiston. The word came from the Greek term for "fire," and the idea was originally proposed by the German alchemist Johann Joachim Becher in the late 1600s and formalized into a theory by his student Georg Ernst Stahl in 1716. According to the theory, things that were flammable contained this invisible substance. When they burned, they released it into the air. They stopped burning when either they ran out of phlogiston or the air became saturated with the substance and couldn't absorb any more. Priestley figured that since things burned longer and hotter in his air, it must be able to absorb more of the stuff than ordinary air. Therefore, he called it dephlogisticated air.

The problem for Priestley, of course, was that phlogiston theory was completely wrong, and there were a number of people at the time, including Lavoisier, who were actively proving as much. According to the theory, when things burned, they were releasing a substance into the atmosphere. If that were true, then after they burned, they should weigh less. Lavoisier, among others, had shown, however, that when materials, such as metals, are heated, their weight increased. Today we know this is due to oxidation. The heated materials are actually combining chemically with oxygen in the air, and that accounts for their increased weight.

Priestley failed to recognize this for a number of reasons. First, the change in weight is small and requires very precise measurements to notice. Second, he was working in a time and place in which science was largely a qualitative enterprise. Priestley, like other English scientists, was used to making observations about qualities: "The flame went out," "There was a disagreeable odor," and "The mouse died." The death of phlogiston theory required the quantitative observations of men like Lavoisier.

Finally, Priestley had the misfortune of being on the wrong side of what Thomas Kuhn called a paradigm shift. Periodically science, like other fields, undergoes a sort of seismic shift when the prevailing theories collapse under the weight of new, contradictory discoveries. This often causes a rift in the scientific community, those willing to abandon old theories and explore new possibilities on one side, and those who hold doggedly to the old doctrine on the other. Paradoxically, Joseph Priestley, who had for so long advocated radical views in areas of religion, education, and politics, now found himself on the orthodox side of the divide.

Once the battle was joined, both sides brought out their big guns. Priestley was by then one of the most respected chemists in the world, and he gave a full-throated defense of phlogiston theory. The challenger, Lavoisier, was unbowed, and he had the empirical evidence to back up his claims. Phlogiston's supporters countered, not without some justification, that Lavoisier had replicated Priestley's experiments and stolen his ideas. The Frenchman's supporters claimed that he had made the discovery first. The debate lasted for months, but eventually, phlogiston and many of its supporters fell by the wayside. Ironically, the theory that replaced it, the caloric theory, would eventually be proven just as wrong, but that would be another battle.

To add insult to injury, the chemical revolution that resulted culminated in the adoption of a system of chemical nomenclature of Lavoisier's design. His new system dispensed with the old chemical names, many of which dated back to the alchemists. He developed a cohesive system for naming chemical elements and compounds that is still in use today. Once it was in place, Priestley's dephlogisticated air was replaced by Lavoisier's now familiar term, oxygen.

In the midst of all this, the war between England and its rebellious American colonies was also under way. Priestley may have been hanging on tenaciously to the conservative side of the phlogiston debate, but he didn't hesitate to voice his support for American independence. He wrote passionately about men's right to liberty and did everything in his power to rally support for his friend Franklin's cause.

He was also not shy about giving his religious views. In 1774, he and his friend the Reverend Theophilus Lindsey founded the Unitarian denomination. Three years later, he published *Disquisitions Relating to Matter and Spirit*. In it and other pamphlets, Priestley questioned the divinity of Jesus Christ, denied the Trinity, and suggested that the concept of a soul might be superfluous. Those in power immediately labeled Priestley a heretic.

Whether these views are what caused the split between Priestley and Shelburne is up for some debate. Shelburne was himself quite liberal, but he was also a man who kept open the possibility of future high office. His continued friendship with Priestley would have made that politically difficult. It may also have been that the two men simply had a falling out. There is also evidence that it was all precipitated by Shelburne's new wife. She seems to have had some antipathy toward both Joseph and Mary. For whatever reason, Priestley and his family moved out of the earl's estate around 1780. It was apparently an amicable split, because Shelburne continued to pay Priestley an annual annuity of 150 pounds.

The family now numbered six, Joseph and Mary; their seventeen-year-old daughter, Sally; and three sons, Joseph Jr. (eleven), William (ten), and little Harry (a toddler). The lot of them moved to Birmingham, where Mary's brother provided them with a comfortable house in Fair Hill. Although less luxurious than what they had become accustomed to, it provided adequate room for the family, as well as a garden, and space for a new laboratory. Priestley wrote numerous accounts of how he enjoyed his time there with his family, and he continued his research. It was in Birmingham that Joseph Priestley would become a full fledged lunatic.

Birmingham was a rapidly growing center for commerce and industry, and it boasted its own equivalent of the Honest Whigs. A number of the leading industrialists and thinkers of the time formed an informal group that included Mary's brother, William Wilkinson, Josiah Wedgwood, Matthew Boulton, whose ironworks were the city's largest employer, the Scottish inventor James Watt, and the scientist, physician, and poet Erasmus Darwin, grandfather of Charles. This illustrious group of men would meet monthly on the Sunday closest to the full moon to make it easier for them to find their way home after a late night of philosophizing and debate. Accordingly, they called the group the Lunar Society.

Priestley was a welcome addition to this band of Lunaticks, as they were commonly called. They even moved their meetings from Sundays to Mondays to better accommodate his ministerial duties. He actively joined in, enjoying the unrestrained exchange of ideas and making his own valuable contributions. Soon, he was acting as a sort of informal science adviser for the group's business leaders, analyzing clay samples for Wedgwood and iron-ore samples for Boulton. He even helped Watt investigate possible improvements for his steam engine. In exchange, the business-men supported Priestley's work financially and supplied him with the finest scientific tools, many of his own design.

As much as Priestley enjoyed his life in Birmingham, all good things must come to an end. The beginning of the end came in 1782, when Priestley decided to follow up on his previous con-troversial work *Disquisitions Relating to Matter and Spirit* with a new book titled *History of the Corruptions of Christianity*. If his earlier work could be said to be incendiary, then the new one was the theological equivalent of the atomic bomb. In it, Priestley gave a historical deconstruction of the Christian church, in which he attempted to delineate every instance of mysticism and super-stition, which he characterized as gross distortions of early Chris-tianity. This included the divinity of Christ, the Holy Trinity, the sanctity of the Eucharist, and what he described as the pagan veneration of saints and angels.

This had the predictable effect of reigniting the charges of heresy. He was uniformly denounced by Anglican Church offi-cials, including Archdeacon Samuel Horsley. Horsley was not only a leader in the church, but also a former secretary of the Royal Society and publisher of Newton's philosophical works. His condemnation carried weight in both the religious and scien-tific communities.

The book was banned in many parts of England, and publicly burned in parts of the Continent. Priestley now found himself shunned by many of his scientific colleagues. It is interesting that a number of copies of the book were sent to America, and one of them found its way into the library of Thomas Jefferson. Unlike many in the Old World, Jefferson praised Priestley's work and would later claim it as one of the primary inspirations for his own *Jefferson Bible*.

The controversy might have passed, as it did after *Disquisitions*, but in 1785 Priestley followed it up by giving what would forever become known as the Gunpowder Sermon. It was November 5, Guy Fawkes Day, commemorating the plot by Catholic conspirators to blow up Parliament in 1605. In the course of his sermon, Priestley used the word *revolution* no fewer than three times and went on to describe the growth of Unitarianism in bold terms:

> Unitarian principles are gaining ground every day. We are, as it were, laying gunpowder, grain by grain, under the old building of error and superstition, which a single spark may hereafter inflame, so as to produce an instantaneous explosion; in consequence of which that edifice, the erection of which has been the work of ages, may be overturned in a moment, and so effectually as that the same foundation can never be built upon again.

Beforehand, a number of Priestley's friends, including Wedgwood, urged him against the use of such fiery language. Whether out of stubbornness, naïveté, or latent self-destructiveness he ignored the advice. He was condemned in speeches and pamphlets and widely lampooned in editorial cartoons. The most famous of these depicts Priestley treading on the Bible while burning various documents related to English freedom. His enemies could now add revolutionary to the charge of heresy, and from that point on, he was saddled with the nickname "Gunpowder Joe."

An ominous pamphlet published under the alias John Nott included the following: "Now, prithee Mr. Priestley, how would you like it yourself, if they were to send you word that they had laid grains of gunpowder under your house or meeting-house?" The none-too-subtle threat was followed in January of 1790 by three men attempting to break into Priestley's home. When they were discovered by his maid, they fired a pistol shot at her and fled.

The only thing that could possibly make things worse was the timing. England was still in the process of recovering from its defeat at the hands of the American rebels, and across the channel in France, the seeds of revolution were quickly taking root. The aristocracy in England and the Royalists who supported them were terrified.

Priestley and his friends on the other hand were delighted. The early stages of the revolution were seen as a logical evolution of the rights of men. The rule of despots seemed to be near its end, and a promised age of equality and reason was at hand. Unable to anticipate the coming horrors of the Reign of Terror, Priestley and his friends voiced wholehearted support for the French Revolution, just as they had done for the revolution in America.

In July 1791, a number of Priestley's friends formed a group calling itself the Constitutional Society. They decided to have a dinner commemorating the fall of the Bastille and ran a notice in the *Birmingham Gazette* announcing their plans to meet at the Royal Hotel. It was followed, shortly thereafter, by another notice threatening to publish an "authentic list of all those who dine at the hotel," signed "Vivant Rex et Regina." Over the course of the following week, a veritable storm of leaflets, handbills, and newspaper notices inflaming passions on both sides ensued.

It is ironic that when the appointed day for the dinner came it was almost called off. Members of the society decided that the situation was just too dangerous and agreed to cancel the whole affair. The owner of the hotel, however, in a last-ditch attempt to save his booking came up with an alternate plan. He suggested that they move the dinner to earlier in the day. That way, they could have their celebration and leave before any trouble started. As an added precaution, several of the members suggested that Priestley, the focus of much of the hostility, remain at home.

At three o'clock, approximately eighty members of the Constitutional Society met at the hotel, had a meal, and made a number of toasts to the French National Assembly, liberty, freedom, and various other liberal causes. By five o'clock, the dinner was over and the attendees departed. The only trouble encountered was a small band of pro-king and -church protesters who pelted them with mud and small stones.

A crisis seemed to have been averted, until around eight o'clock, when a large group of protesters arrived from the local pubs intent on giving a piece of their minds to the revolutionary sympathizers. When they were informed that they had missed the targets of their ire by a number of hours, they became enraged. They rioted and broke a number of windows at the hotel, but when that proved insufficient to satisfy their fury, the crowd moved on to Priestley's New Meeting House.

In an orgy of misdirected passion, the drunken rioters demolished and burned the entire building. When that was accomplished, a group of them moved on to exact a similar vengeance on the Old Meeting House. After that, they marched to Priestley's home in Fair Hill. Having witnessed much of this, Priestley's friend Samuel Ryland raced ahead of the mob to warn him. At first, Priestley refused to believe the reports of such violence, but Ryland eventually convinced him and Mary to flee.

They grabbed what few things they could and barely managed to leave before the mob arrived. From the relative safety of their friend William Russell's house, Joseph and Mary could see the flames of the burning meeting houses. When the rioters arrived at Fair Hill, Priestley's twenty-year-old son, with Russell's help tried to calm the mob. Eventually, however, they had to flee themselves as Priestley's home and laboratory were burned to the ground. Countless irreplaceable documents and scientific instruments were destroyed as well as much of Priestley's faith in his fellow man.

Even then, the drunken mob's thirst for vengeance was unsatisfied. They got wind that Priestley was at Russell's and pursued their prey. What followed was a desperate chase with the Priestleys rushing from one temporary sanctuary to the next, until eventually they were forced to leave Birmingham altogether. For four days, the mob continued its rampage until the king was reluctantly forced to send in troops to put down the disturbance. By the time it was over, the Birmingham riots cost the lives of scores of rioters and more than a dozen homes and churches, including the homes of Russell and Ryland.

Priestley made it to London, where he remained underground for weeks. Even with the support of his friends in the Lunar Society, though, his days in his native England were numbered. The final catalyst for him leaving the country might have been the news that the revolution in France had gotten so out of hand, that Lavoisier, France's most prominent chemist and Priestley's old nemesis, had met his end on the guillotine. In the spring of 1794, Joseph and Mary Priestley set sail on board the *H.M.S. Samson* for America. Even though his long-time friend Benjamin Franklin had died a few years before, Priestley hoped that the new nation would provide a more welcoming attitude toward dissenters. He, Mary, and their sons settled in Northumberland in rural Pennsylvania, where he would spend the remainder of his days.

It was a remarkable journey. Joseph Priestley had begun in obscurity as an unconventional pastor, become a teacher and advocate of educational reform, founded a new church, formed friendships with some of the greatest scientists of the day, and risen to become one of them himself. He endured controversies, ridicule, and scorn and made scientific discoveries the significance of which would not be fully appreciated for almost two centuries. In the course of one of those controversies, he made what would ultimately be his most suitable epitaph. In response to one of his critics, he wrote:

> It may be my fate to be a kind of comet, or flaming meteor in science, in the regions of which . . . I made my appearance very lately, and very unexpectedly; and therefore, like a meteor, it may be my destiny to move very swiftly, burn away with great heat and violence, and become as suddenly extinct (Jackson, 2005).

As the eighteenth century gave way to the nineteenth, the chemical revolution that Priestley and Lavoisier had ignited was burning brightly. Every day it seemed scientists were discovering new elements, which they named using Lavoisier's system. Priestley's pneumatic chemistry was unlocking more and more secrets of the air, and the knowledge of chemists was being seen as a valuable commodity by the captains of industry.

All this set the stage for the rise of a new star of chemistry. He was a young man of rural origins, similar to his inspiration, Priestley, and he would go on to electrify the world of chemistry. Along the way, he would inspire both adulation and controversy, and would be remembered as the scientist with the soul of a poet. His name was Humphry Davy.

In the far southwest corner of Britain sit the moors and windswept cliffs of Cornwall. Into that untamed landscape in December 1778 Humphry Davy was born. His father, Robert Davy, was a woodcarver from an old Cornish family. Although talented and ambitious, he had a series of misfortunes and a history of gambling. He died when Humphry was fourteen, leaving the family seriously in debt.

Unlike many of the figures we've examined, Humphry Davy, the oldest of five children, had a long and close relationship with his mother, Grace. She and his grandmother doted on the boy and instilled in him an early love of reading. As a young child, he would entertain friends and family by reciting passages of his favorite books or poems of his own creation. Even then, the boy seemed to revel in the attention of an audience.

Although he displayed a fine intellect, he was an undisciplined child, who spent much of his youth hunting, fishing, and exploring the Cornish wilds. This gave rise to a lifelong love of nature and an appreciation of its beauty. Although he attended school, he failed to pick up much of the classical education given, and what he did learn was the result of his own reading and investigation. He did show an early interest in chemistry, but it seems to have taken the form of making his own homemade fireworks, his younger sister, Kitty, acting as his first lab assistant.

Shortly after his father died, Davy was apprenticed to a local apothecary-surgeon named John Bingham Borlase. The apprenticeship gave Davy a chance to explore chemistry in a less dangerous direction and learn the proper tools of a scientist. To help, Borlase gave Davy access to his library and allowed him to read texts on chemistry, including Lavoisier's *Elements of Chemistry*.

His self-education was aided by his friendship with local saddler and instrument maker Robert Dunkin. Dunkin was a man of wide interests, including science. He constructed various scientific tools for himself, such as Leyden jars and machines similar to the ones Priestley used for generating electrical sparks. Together Davy and Dunkin conducted their own amateur experiments.

A number of these concerned heat and light, and their results contradicted much of the caloric theory Lavoisier had used to bludgeon Priestley's phlogiston. Davy became convinced that contrary to Lavoisier's idea that heat was a type of invisible fluid, it was actually the result of movement. This was much closer to the modern view than either phlogiston or caloric theories.

During this period, Davy also began seriously composing his own poetry. Many of his early poems centered on his beloved outdoors and the beauty of nature. One, however, gave some indication of the direction his life would take. Titled "The Sons of Genius," it contained the following stanzas:

To scan the laws of nature, to explore
The tranquil reign of mild Philosophy;
Or on Newtonian wings to soar
Through the bright regions of the starry sky

From these pursuits the sons of genius scan
The end of their creation,—hence they know
The fair, sublime, immortal hopes of man
From whence alone undying pleasures flow.

Theirs is the glory of a lasting name,
The meed of genius, and her living fire;
Theirs is the laurel of eternal fame,
And theirs the sweetness of the muses lyre (Knight, 1992).

Just as Priestley's career in science was influenced by lucky meetings, so too was Davy's. To make ends meet after the death of her husband, Davy's mother took in boarders. One of these was Gregory Watt, son of James Watt of steam engine and Lunar Society renown. The younger Watt suffered from tuberculosis and had been sent to Cornwall to recuperate. He and Davy became great friends. Both young men were interested in science, and Watt, who had been educated in Glasgow, shared his knowledge of chemistry with his new companion.

Through Watt, Davy also made other useful connections, including Thomas Wedgwood, son of Josiah Wedgwood. In addition to these Lunatick contacts, Davy was also introduced to Davies Gilbert, the sheriff of Cornwall. He was an educated man, well versed in science, and he recognized Davy's talent. He encouraged the boy, gave him access to his extensive library, and used his influence to further Davy's career.

Gilbert also introduced him to Thomas Beddoes. Beddoes was a chemistry lecturer at Oxford, who also had connections to the Lunar Society and, like Priestley, had expressed sympathy for the French Revolution. This made him a political outsider and destroyed any chance he had of becoming a full professor. Instead, he had plans for a medical facility in Bristol to take advantage of the "factitious airs" that Priestley had discovered. The Pneumatic Institute was to investigate and promote the use of gasses, such as oxygen, as a means of treating disease.

When Beddoes met Davy he was so impressed that he offered to publish the findings of his experiments. Davy's "Essay on Heat and Light" was included in 1799 as part of a symposium Beddoes edited, *Contributions to Physical and Medical Knowledge, Primarily from the West of England*. Although the essay had serious flaws, reflecting Davy's lack of formal experience, it demonstrated the freshness of his ideas. When Joseph Priestley, by then in America, read the work, he wrote, "Mr. H. Davy's Essays . . . impressed me with a high opinion of his philosophical acumen" (Terneer). It also happened that Beddoes needed someone to run his Pneumatic Institute, and on Gilbert's recommendation, he offered the twenty-year-old Davy the job.

Beddoes was not only a man of science but also a friend of the literary world. Once Davy had moved to Bristol, Beddoes eagerly introduced the young man to his publisher and writer friends, and Davy, the aspiring poet, was delighted to be brushing elbows with the Bristol literati. The publisher Joseph Cottle described his first meeting with Davy:

I was much struck with the intellectual character of his face. His eye was piercing, and when not engaged in converse, was remarkably introverted, amounting to absence, as though his mind had been pursuing some severe trains of thought, scarcely to be interrupted by external objects; and from the first interview also, his ingenuousness impressed me as much as his mental superiority.

Cottle was publisher of the Lake poets Robert Southey, Samuel Taylor Coleridge, and William Wordsworth, and it wouldn't be long before Davy made their acquaintance as well.

Once the Pneumatic Institute was established, Davy set to work investigating the properties of gasses. As part of those investigations, he decided to follow Priestley's example and inhale them himself to witness their effects. Fortunately for Davy, one of the first gasses he chose was one Priestley had discovered, nitrous oxide.

Priestley had called this dephlogisticated nitrous gas, but little was known about it at the time. The American physician Samuel Mitchill had proposed that it was extremely harmful and may actually be the cause of disease. He called it oxide of septon and maintained it would be fatal if breathed and was capable of causing wounds if it even came in contact with skin. Davy knew that the claims of the gas's corrosive effects on skin were false because he had exposed animal carcasses to it with no effect. With that in mind, and with the aid of a colleague, Dr. Kinglake, he filled a green silk bag with Priestley's gas, put it over his mouth and nose and breathed deeply. He later reported what happened:

A thrilling, extending from the chest to the extremities, was almost immediately produced. I felt a sense of tangible extension highly pleasurable in every limb; my visual impressions were dazzling, and apparently magnified. . . . By degrees, as the pleasurable sensation increased, I lost all connection with external things. . . . I existed in a world of newly connected and newly modified ideas. I theorized; I imagined that I made discoveries. I was awakened from this semi-delirious trance by Dr. Kinglake, who took the bag from my mouth, indignation and pride were the first feelings produced by the sight of the persons about me (Terneer, 1963).

In the spirit of scientific thoroughness, Davy decided to repeat the experiment the following day. The results were equally pleasurable. What's more, Davy made the delightful discovery that, unlike strong drink, the wondrous new gas produced its desired effects without producing a hangover. Kinglake tried the gas and confirmed his results. Davy then did what any serious young scientist would do in a similar situation. He immediately began telling his friends and colleagues about a great method he had discovered for getting high.

Soon, nitrous oxide was being tried by other researchers and earning the nickname Davy gave it, laughing gas. Among the medical professionals who experienced its pleasurable effects were Beddoes and Peter Roget, of thesaurus fame. Davy began using it recreationally, sometimes as frequently as three or four times a day. Although he eventually weaned himself off, Davy introduced it to his friends, including Southey and Coleridge. After sampling the gas, Coleridge described that he, "could not avoid, nor indeed felt any wish to avoid, beating the ground with my feet; and after the mouth-piece was removed, I remained for a few seconds motionless, in great extacy."

Word spread, and for a while nitrous oxide parties were all the rage among England's high society. Davy essentially became the Timothy Leary of his day. He suggested that the gas might be useful for dulling pain during operations, but it wasn't used for that purpose until the 1840s. In 1799, Davy published his experiences with the gas in *Researches, Chemical and Philosophical; Chiefly Concerning Nitrous Oxide, of Dephlogisticated Nitrous Air, and Its Respiration*. The book became a scientific hit and Davy's reputation grew.

Buoyed by this success, Davy unfortunately tried inhaling other gases. He tried nitric oxide, which unlike its nitrous cousin forms nitric acid when combined with water. When Davy inhaled it, the moist air in his breath caused the acid to form, and he badly burned his throat, esophagus, and parts of his lungs. Once he recovered from that, he tried the same thing with something called water gas, a combination of carbon monoxide and hydrogen, that very nearly killed him. As a result of these and similar injuries, Davy developed a series of respiratory problems and spent the later part of his life as an invalid.

In spite of Davy's rising fame and the popularity of nitrous oxide, the Pneumatic Institute began to flounder. Beddoes wasn't able to effect the cures he hoped for. Many of the locals considered some of his treatments, such as housing patients in the lofts of barns so they could enjoy the healthful effect of the methane produced by the cows, to be little more than quackery. The supply of patients dried up, and Davy's prospects for continued employment were in doubt. Fortunately, his earlier successes brought him to the attention of Count Rumford, who offered him a job at a newly formed institute dedicated to the practical use of science. It was located in London, modeled on the Conservatory of Arts and Trades in France, and came to be called the Royal Institution.

In its early days, the Royal Institution was in search of wealthy patrons. Fortunately, Davy showed a natural aptitude not only for chemistry but for fund-raising and publicity as well. He began delivering public lectures on science. These presentations took advantage of his good looks, charisma, and way with words, and they proved both entertaining and popular. As the dashing young Davy took the stage, he waxed eloquent about the wondrous discoveries of chemistry, accompanied by spectacular demonstrations of his own design. Wealthy patrons, not a few of them female, flocked to the lectures and not only supported the institution but made Davy a star.

It didn't hurt that Davy's first lecture was about nitrous oxide, but he soon began lecturing on an equally exciting topic, electricity. The Italian scientist, Alessandro Volta, had created the first battery in 1800. Called an electric pile, it consisted of copper and zinc plates separated by disks of cardboard moistened with saltwater. Upon reading about the device, Davy and many other chemists became fascinated with the potential for using it to unlock the secrets of matter.

Davy began experimenting with his own version of Volta's pile while still at the Pneumatic Institute. He found he could improve on Volta's design by doing away with the cardboard disks and immersing the metal plates in acid. Then Davy showed that he could break down salts and acids into their components by passing an electric current through them. He demonstrated this for the public in one of his Royal Institution lectures, founding what would become known as the field of electrochemistry. So significant was Davy's discovery that, despite England and France being at war at the time, Napoleon awarded him a medal and a three-hundred-franc prize for his groundbreaking work with electricity.

In 1803, Davy was elected to the Royal Society, and became one of its secretaries in 1807. With the power of electricity now at his disposal, he turned his attention to a material that had previously defied all attempts at analysis, potash. Now known as potassium hydroxide, potash was derived from wood ashes, and at first, it resisted Davy's attempts as well. Solid potash doesn't conduct electricity, and when he tried dissolving it in water, the electric current succeeded only in breaking apart the water molecules into hydrogen and oxygen. Davy then tried melting it, eventually using the heat of the electric arc to produce the molten potash. This led to a spectacular success.

As Davy looked on, the electricity began to liberate small globules of a shiny metal from the molten liquid. They bobbed to the surface, reacting with the water present to produce hydrogen, which the heat of the reaction immediately ignited. Before his eyes, Davy saw the bits of unknown metal burst into lavender flame. He had discovered the element potassium. Edmund Davy, Humphry's cousin, was present, acting as lab assistant, and he described his famous cousin's reaction:

> When he saw the minute globules of potassium burst through the crust of potash, and take fire . . . he could not contain his joy—he actually danced about the room in ecstatic delight; some little time was required for him to compose himself to continue the experiment (Solomon, 1973).

Davy went on a week later to use the same method to discover the metal sodium from caustic soda with similar fiery results. Over the course of the following year, he went on to discover five more elements: barium, calcium, boron, strontium, and magnesium. You can imagine the reaction as he demonstrated his results before live audiences as part of his lectures. As amazing as his laboratory discoveries were, it was the lectures themselves that led directly to what many consider to be Davy's most significant discovery, Michael Faraday.

Faraday was the son of a London blacksmith, and as a young man, became apprenticed to a bookbinder. While learning how to bind books, the curious young man took the liberty of reading them, and quickly became fascinated by the ones concerning science, in particular chemistry. As luck would have it in 1810, one of the bookbinder's customers gave Faraday a ticket to hear one of Davy's lectures.

Davy was little more than a decade older than Faraday and at the height of his success. As soon as he began his lecture with characteristic flare, the young bookbinder's assistant became enthralled. He took notes on the lecture, which he later bound into a manuscript with illustrations and sent to Davy, along with an inquiry about a job.

By the time the manuscript arrived, Davy was conducting research into a newly discovered compound called nitrogen trichloride. It's an extremely unstable material, and its discovery cost the chemist Pierre Louis Dulong an eye and a finger in the resultant explosion. Davy initially had similar results, but fortunately the accidental detonation only caused him temporary blindness. Davy continued his work, but now needed someone to transcribe his notes for him. Faraday fit the bill nicely and performed so well that, later, when Davy needed a new lab assistant he gave Faraday the job.

Around this time, Davy undertook one of his most perilous and least successful experiments. He got married. Despite the attention of his numerous female fans, Davy remained unattached for many years, but in his thirties he fell in love with a wealthy Scottish widow, Jane Apreece. She was an intelligent and beautiful young woman from an aristocratic family, who often hosted her fellow female intellectuals at one of the discussion groups known as Bluestocking clubs.

In April 1812, Davy was knighted and became Sir Humphry Davy. A few days later, he became a married man. Unfortunately, the reaction between two such passionate and strong willed individuals proved to be rather combustible. They argued frequently, and even though they remained married and maintained a genuine respect for each other, they eventually ended up living apart.

However, in 1813, Davy traveled to France with his bride to accept his prize from Napoleon. Faraday accompanied them as Davy's secretary and valet. Although Davy appreciated the younger man's company and treated him as a close assistant, the new Lady Davy seems to have treated Faraday as just another one of the servants.

Faraday continued to serve as Davy's assistant for many years. In return, Davy acted as Faraday's mentor, furthering his scientific career. Alas, the relationship between patron and protégé, like that between father and son, can often become strained, and that's what happened between the two men. Faraday went on to become a great scientist and make significant chemical and electrical discoveries of his own. Davy became jealous as his former assistant's fame began eclipsing his own. In 1823, when Faraday was nominated for membership in the Royal Society, Davy, then the society's president, opposed the nomination. Faraday's nomination succeeded, despite these objections, but the relationship between the two would remain cool from then on.

As Faraday's star continued to rise, the years of chemical exposure and accidents began to take a toll on Davy's health. When he was only in his late forties, he suffered from heart trouble and had an apparent stroke. When he was reelected as president of the Royal Society he gave an address at the banquet. His brother, John, who was present for the occasion described his appearance:

> When he delivered that discourse which was his last to the Royal Society, at the anniversary meeting on St. Andrew's day, 1826, it was done with such effort that drops of sweat flowed down his countenance, and those near him were apprehensive of his having an apoplectic seizure; and he was so much indisposed after, that he was unable to attend the dinner of the Society (J. Davy, 1858).

Shortly after that, he traveled to the Continent to convalesce and would stay there for the remainder of his life. He died in a hotel in Geneva in 1829, at the age of fifty. It's interesting to note that early in his career Davy visited the home of writer William Godwin. While there, he met Godwin's young daughter Mary. Years later, as Mary Wollstonecraft Shelley, she would create a character in her novel, Frankenstein that bore more than a passing resemblance to the young Davy. In the novel, Victor Frankenstein attends a lecture by a charismatic chemist named Waldman. He becomes captivated, much as Faraday did, by the lecturer's passionate description of chemical inquiry and the pursuit of knowledge. Davy's own lectures were no less inspirational, and in one in particular he told his awed listeners:

> Nothing is so fatal to the progress of the human mind as to suppose that our views of science are ultimate; that there are no mysteries in nature; that our triumphs are complete, and that there are no new worlds to conquer.

CHAPTER FOUR

The Bloody Path

From Grave Robbing to Modern Surgery

THE ALCHEMIST AND THE CHEMIST WERE BOTH CAPABLE of inspiring fear of the mad scientist, as they bent over their glassware and crucibles, delving into the secrets of matter, but a figure capable of tapping into a far more primal fear was the bloody-handed surgeon. From ancient times, surgery was at best a brutal art. Before the advent of anesthesia and antiseptic techniques, the surgeon was as likely to kill as cure. Add to that the lowly status accorded surgeons by physicians. The word *surgeon* comes from the Greek words meaning "hand work."

That status began changing in the 1700s, due, in no small part, to the efforts of a maverick young Scotsman. Over the course of his storied career, he rose from the rank of procurer of bodies and part-time grave robber to become one of England's most celebrated surgeons and anatomists. Along the way, he earned both enthusiastic disciples and powerful enemies. His collection of anatomical specimens and a menagerie of exotic beasts and monsters both titillated and terrified Georgian London. His teachings influenced generations of surgeons, and he would be responsible for saving millions of lives. His name was John Hunter.

In a two bedroom cottage in Long Calderwood, Scotland, south of Glasgow, John Hunter was born on February 14, 1728, the youngest of ten children. Three of his older siblings, including one also named John, died before he was born. His father, John Hunter Sr., a farmer and grain merchant, was sixty-five at the time his youngest son entered the world and seems to have focused most of his attention on the boy's two older brothers, James and William. Raising young John, or Jock as he was nicknamed, fell to his mother, Agnes, and his four older sisters.

From an early age, he was a headstrong child who, like Davy, eschewed formal education in favor of exploring the fields and forests. As he described his school experience:

> When I was a boy it was a little reading and writing, a great deal of spelling and figures; geography which never got beyond the dullest statistics, and a little philosophy and chemistry as dry as sawdust, and as valuable for deading [deadening] purposes. I wanted to know about the clouds and the grasses, why the leaves change colour in the autumn. I watched the ants, bees, birds, tadpoles, and caddis worms. I pestered people with questions about what nobody knew or cared anything about (Paget, 1897).

This was a sharp contrast to his older brothers who excelled at school. James, the eldest, and William both went to Edinburgh University and studied medicine. Upon graduation, they traveled to London to further their careers. Unfortunately, plans they had to set up a joint practice were shattered when James contracted tuberculosis and died. Despite this tragedy, William met with success. He received his license as a surgeon, establishing himself as a man midwife, what we would call an obstetrician, attending the births of the well to do. He also made ends meet by establishing a private anatomy school.

Meanwhile, young John dropped out of school. The closest thing he had to a career was assisting his brother-in-law as a carpenter. With no other prospects, he toyed with the idea of joining the army, but instead decided to write to his brother for advice. The timing of that letter would change John's life and eventually usher in a new age in surgery.

William was in desperate need of an assistant to help with his rapidly growing school, and John's letter came shortly before the start of the fall term. William urged his younger brother to come to London. When John arrived in 1748, at the age of twenty, William welcomed him and made him his anatomical assistant. Whether because it took advantage of his experience with carpenter's tools or because it was an extension of his curiosity into the wonders of nature, John took to his new craft with relish. As he bent over the dissecting table, teasing apart muscles, veins, and tendons from human cadavers, John showed an aptitude that impressed and delighted William. The older brother soon encouraged the younger and remarked that he would make a fine surgeon someday.

The problem was that the cadavers John was practicing on

and that William needed for his school, were in short supply. That's where the primary part of John's job came in. Medical professionals at the time in England were allowed to use the bodies of executed criminals for educational dissections, but despite the frequency of executions, the demands of all of the medical students in London soon outstripped the limited supply. Into that void entered the grave robbers. Often referred to as resurrectionists, these criminal entrepreneurs where ready and able to supply the goods, for a fee.

William had ambitions of upward mobility, taking pains to soften his Scottish burr and dress in the manor of a gentleman. Associating with the criminal element would have undermined the refined image he cultivated. As a roughhewn rural youth with a love of taverns and a propensity toward frequent cursing, John had no such qualms. He quickly made the acquaintance of a number of grave robbers and was soon providing them with plenty of work. He spent many nights by the school's back door overseeing their covert deliveries and sometimes even accompanied them on their nocturnal excavations.

With this new supply of learning materials at his disposal, John had plenty of time to explore the intricacies of the human body. Over the next few years, by his own tally, he performed over a thousand dissections, providing him with a far better medical education than most physicians of the day. Although increasingly frequent on the European Continent, dissection of human cadavers was still rare in Great Britain. Traditionally, they were performed for an audience of medical students by a professor, who stood on a pedestal and read from an accepted text about what they were supposed to see. Meanwhile, his assistant would briefly expose parts of the body, being careful not to contradict what the professor said. Students watched from a distance without benefit of actually touching the corpse. William's genius was to see that would-be medical men could learn far more hands on, and he set up his anatomy school for just that purpose.

Under William's tutelage, John also learned the finer points of preparing anatomical specimens. Before the advent of plastic models, carefully prepared specimens where used for teaching purposes, and John eagerly took up the craft of preparing them. He practiced his new art, drying bones and strips of muscle, injecting blood vessels with wax and pickling various organs in preservative spirits, proving himself so adept that a number of his preparations are still on display in medical museums.

Between his daylight dissections and nighttime forays, John found time to faithfully attend his brother's lectures. William was a fine lecturer, and between what John learned in class and the skills he honed with his dissecting knives, within a few months, he was instructing the other students. His older brother was so impressed that by the spring of 1749, he pulled a few strings and arranged for John to study with William Cheselden, one of the most respected surgeons in England.

This was quite a coup. Cheselden was a giant in the surgical field. He was head of the Company of Surgeons, and his book *The Anatomy of the Human Body* and his atlas of the skeletal system, *Osteographia,* were considered must-haves for medical students. Although Cheselden was in semiretirement, the chance to walk the wards of Royal Hospital in his shadow was a golden opportunity for John. Not only did it provide a quicker path to becoming a surgeon than the usual lengthy apprenticeship but it allowed young Hunter to learn from someone not so bound by tradition.

Most surgeons in the eighteenth century relied on the same classical texts that Paracelsus had railed against centuries before. Their standard techniques included purging and bleeding to balance the four humors Galen had written about. In contrast, Cheselden was willing to use observation and experimentation, to learn from his experiences and modify his practice as needed. What's more, he viewed surgery as a last resort, not undertaken lightly. Hunter learned these lessons well.

For two years, John spent his summers at the hospital with Cheselden and his winters bent over the dissecting tables at William's school. In 1751 Cheselden fell ill, and died, and John's education fell to an up-and-coming surgeon named Percivall Pott. He followed many of Cheselden's teachings, including his reluctance to perform avoidable surgeries.

By 1754, John was ready to apply the extensive knowledge he got from the dead to help the living. He became a surgical student at St. George's Hospital and began treating his first patients. Many of the cases saw him applying the lessons learned at Cheselden's side. He observed the outcomes of traditional treatments, formed hypotheses about ways to improve treatments, conducted experiments, and implemented the improvements. Considered common practice now, many of his surgical colleagues saw this radical approach as nigh onto heresy.

While John and his patients enjoyed his newfound success, tensions began building between the Hunter brothers. The increasing demands of William's flourishing practice were taking him away from the school, and John was forced to take up the slack. Not only was he instructing students in dissection but he was also trying to conduct his own research and delivering many of William's lectures.

To make matters worse, in the course of his own research, John made a number of significant discoveries that William, as head of the school, took credit for. This included a series of painstakingly assembled preparations demonstrating the blood supply of the fetus and the pregnant mother were separate rather than joined. Not only did William take credit for the discovery but claimed ownership of the preparations that had taken John months to assemble. John was outraged but had little recourse.

By 1760, John had enough. He quit his brother's employ and joined the army. Hippocrates the father of medicine said, "He who wishes to be a surgeon should go to war." England was in the fourth year of what became known as the Seven Years' War, and John Hunter heeded the advice. He accompanied British troops as they stormed the French beaches of Belle Île, an island south of Brittany. While the battle raged, Hunter dug musket balls and shrapnel from torn flesh and shattered bones. He amputated limbs, drained infected wounds, and tended as best he could to the dying, all under horrific conditions like he had never seen during his hospital days.

Once a position on the island had been secured, make-shift field hospitals were set up. By the time the fighting was done, the number of British soldiers killed or injured climbed to over seven hundred. As Hunter and his comrades tried to recover, they were attacked by wave after wave of infectious disease. Typhus, dysentery, malaria, and smallpox collectively took a greater toll than all the enemy fire. In the midst of these hellish conditions, Hunter tried to apply the lessons he had learned from Cheselden.

Standard treatment for gunshot wounds was for the surgeon to open the wound, called dilation, pry out the musket ball or shot with forceps (or, if need be, the fingers), and remove as much foreign material and debris as possible. The theory was sound, but in the filthy conditions of the battlefield it often did more harm than good. The wounds almost always became infected and gangrene quickly set in.

By chance, Hunter had an opportunity to see an alternative. In the course of the fighting, a group of five French soldiers had been shot, but managed to escape. Although four of the five had been seriously wounded, one shot in the arm, two shot in the legs (one with a musket ball lodged in his thighbone), and one shot in the chest, they managed to make it to an abandoned farmhouse where they hid for four days. Eventually, they were captured, but when Hunter examined their wounds, he was surprised to see that in spite of their not receiving any medical care, their gunshot wounds were healing better than the wounds of British patients who had received treatment.

Hunter took this lesson to heart and began advocating for more conservative treatment. Thirty years later, he used the notes from the assault on Belle Île as the basis for his *Treatise on the Blood, Inflammation and Gun-shot Wounds*. In it, he praised the healing powers of nature and explained, "No wound, let it be ever so small, should be made larger, excepting when preparatory to something else" (Palmer, 1835).

By early 1762, things had quieted on Belle Île. The wounded had been evacuated, and Hunter was appointed chief surgeon and director of hospitals on the island. Most soldiers were reassigned to duties elsewhere, and all that remained was a small defensive contingent. This left Hunter plenty of time to indulge his childhood pursuit of nature. He spent most of his days hiking or riding around the island making observations about the sea birds and other local wildlife.

Hunter had been interested in comparative anatomy, and often used animals to investigate features difficult to see in humans. For instance, while still in London, he conducted an experiment on a dog in which he opened up the living animal's abdomen, and poured milk through a hole in its intestine to demonstrate the role of the lymphatic system in the absorption of fat. While horrifying by today's standards, such vivisection was common in the eighteenth century and vital to understanding living systems.

While on the island, Hunter began collecting local lizards to study the little understood phenomenon of hibernation. He collected the slumbering lizards and force fed them to see the action of their digestive systems while they hibernated. In the course of his studies, he made a remarkable discovery, finding that if he grabbed the lizard's tail, it would frequently come off in his hands, allowing the lizard to escape. Not only would the tail continue to writhe for some time once detached, but the lizard itself was able to regrow a new one.

Hunter retained his fascination with regeneration for the remainder of his career. He was particularly interested in the lizards that instead of simply regrowing one new tail would grow two. Such oddities of nature intrigued him. When he was later transferred to Portugal, for the remainder of the war, he carried with him quite a collection of such lizards, carefully preserved.

When the war ended, Hunter returned to civilian life, but without a job. In John's absence, William found a new anatomy assistant, filling his former job at the school. Hospital positions were quickly filled by those with better connections, and his continuing quarrels with his brother made a joint practice impossible. The solution was to form a partnership with a fellow Scot named James Spence. He wasn't a surgeon, but he was more than willing to take advantage of Hunter's surgical knowledge to expand his dental practice.

In the medical hierarchy, dentists were considered a rung down from even the lowly surgeons, but the recent introduction of cheap sugar to the English diet and the subsequent effects on English teeth were making their skills increasingly valuable. Owing to his years of dissection, Hunter had greater knowledge of teeth and jaws than any other surgeon in England. Spence brought his own substantial knowledge to the practice, eventually earning an appointment as "operator for the teeth" for George III. For the next five years, the partnership of Spence and Hunter not only earned them renown among their wealthy clients but also furthered the science of dentistry.

Hunter applied his scientific methods to the discipline of dentistry. He produced two major research works on teeth and jaws: *The Natural History of the Human Teeth*, published in 1771, and *A Practical Treatise on the Diseases of the Teeth*, published the following year. They caused many future dentists to credit Hunter as being the father of scientific dentistry and give the teeth the scientific names they're known by today: incisors, cuspids (canines), bicuspids (premolars), and molars. On a more macabre note, Hunter combined his growing dental knowledge with his fascination for regeneration and used them for the basis of a bizarre series of experiments.

As he watched Spence relieving the suffering of his patients by yanking out rotted teeth, Hunter's mind returned to the lizards of Belle Île and their remarkable ability to regenerate. What if teeth could do the same thing? With that in mind, he began an unusual series of grafting experiments. First he removed a spur from the foot of a young rooster and surgically attached it to the bird's comb. Amazingly the spur appeared healthy and continued to grow. Encouraged, Hunter next removed a testis from another rooster, and implanted it into the bird's belly. Again, the transplanted tissue continued to grow. At that point, he decided to see what would happen if he transplanted tissue from one individual to another. He removed a testis from a third rooster and grafted it into the belly of a hen. Incredibly, this too was successful.

Thrilled with his results, Hunter then attempted one of the earliest examples of cross-species transplantations. He took a healthy human tooth, "donated" by one of London's ubiquitous beggars or street urchins for an undisclosed sum, and grafted it onto the comb of one of his roosters.

I took a sound tooth from a person's head; then made a pretty deep wound with a lancet into the thick part of a cock's comb, and pressed the fang of the tooth into this wound, and fastened it with threads passed through other parts of the comb. The cock was killed some months after, and I injected the head with a very minute injection; the comb was then taken off and put into a weak acid, and the tooth being softened by this means, I slit the comb and the tooth into two halves, in the long direction of the tooth. I found the vessels of the tooth well injected, and also observed that the external surface of the tooth adhered everywhere to the comb by vessels, similar to the union of a tooth with gum and sockets (Moore, 2005).

Unfortunately, Hunter was wrong. He thought the tooth had been successfully transplanted, but unbeknownst to him, there was no blood supply to the tooth. It was dead. Undeterred, he preserved the cockscomb with its attached tooth, and the specimen remains to this day in the Hunter museum.

Having satisfied himself of a transplanted tooth's ability to survive and flourish, Hunter began performing tooth transplants on human patients. He promoted it as an alternative to the unsightly and uncomfortable false teeth available at the time. It continued to be practiced until well into the next century, when concerns about safety and ethical considerations forced it out of fashion.

While assisting Spence and conducting experiments, Hunter continued to see patients. One of his old army comrades, a fellow surgeon named Robert Home, asked Hunter to stop by his house and offer a second opinion for one of his patients. It turned out that the patient was Home's daughter, Anne. The young woman was suffering from numbness on her left side, and developed a series of puzzling blue spots on her leg. Hunter discovered that she was also experiencing bouts of what were described as "Wind in her stomach, especially at night" (Dobson, 1968). He put the clues together, and diagnosed her as suffering from one of the intestinal parasites all too common in the period, and prescribed a course of worming medicine and purgings.

Anne made a full recovery, and in the course of the treatment, Hunter and his patient fell in love. It was an odd match. Anne was tall and beautiful, with blond hair, blue eyes, and fair complexion. She was intelligent, highly polished, and an accomplished poet, who delighted in the sparkling conversation of the Bluestocking clubs. John, on the other hand, was a gruff little man of rough manners, who would rather carve up dead bodies than crack open a book. As ill-matched as they appeared, they were engaged within the year, and went on to have a long and happy marriage. The actual wedding, however, would have to wait almost seven years, while Hunter secured his fortunes and conducted his most dangerous experiment.

As Hunter's clientele grew, so too did the number of cases experiencing the common scourge of the time, sexually transmitted disease. Gonorrhea and syphilis were running rampant through the population, driven by the period's relatively lax attitudes about sex and the growing abundance of prostitutes. Every stratum of society was affected, and Hunter's list of venereal clients included chimney sweeps, prostitutes, soldiers, servants, lords, ladies, vicars, and, in at least one case, a Prussian count. To make matters worse, neither disease was well understood.

Each is caused by a separate bacteria, but doctors at the time believed they were simply different stages of the same disease. If the symptoms were localized to the genitals, then it was said to be gonorrhea. The less localized symptoms of syphilis were attributed to gonorrhea that had spread systemwide. Hunter wanted to conduct an experiment to settle the matter once and for all. He immediately ran into two problems. First, he didn't have a suitable animal to conduct the experiments on, and second, if he decided to experiment on humans, he needed to find a human volunteer willing to be infected. He solved the dilemma by taking what he considered the only ethical course, and conducted the experiment on himself.

In the spring of 1767, he dipped one of his surgical instruments into the open sore of an infected prostitute, and proceeded to jab his own penis with the infectious matter. He then calmly recorded the experiment: "Two punctures were made on the penis with a lancet dipped in venereal matter from a gonorrhoea;

one puncture was on the glans, the other on the prepuce. This was on a Friday; on the Sunday following there was a teasing itching in those parts, which lasted till the Tuesday following" (Palmer).

Unfortunately for Hunter and contrary to medical doctrine of the time, which maintained an organ couldn't be infected by two separated diseases at the same time, the prostitute's sore harbored both the bacteria for gonorrhea and syphilis. Hunter contracted both. He dutifully recorded his symptoms, including swelling, discharge, and pain when urinating and shortly thereafter, began to show syphilitic symptoms as well, such as ulcers on the foreskin, swelling of the lymph nodes, and eventually swelling and soreness on the tonsils.

Hunter treated himself with Paracelsus' old standby, mercury, and eventually the symptoms went away, possibly due to the mercury, but more likely the syphilis simply entered its secondary noninfectious form. Hunter considered the experiment a success, and in 1786 published his results. He reaffirmed the mistaken notion that the two diseases were one and the same, and set back the study of sexually transmitted diseases by fifty years. It wouldn't be until 1838 that the French/American physician Philippe Ricord would use his own controversial experiments to show they were two separate afflictions.

Hunter also continued his work as a comparative anatomist. In addition to human cadavers, he performed numerous dissections on all manner of animal species. As well as his beloved lizards, he took copious notes of the anatomy of fish, turtles, poultry, cats, dogs, donkeys, and various livestock. When that failed to satisfy his curiosity, he began searching far and wide for additional animals to add to his growing notebooks, and was soon making contacts with wild animal dealers, circus and side-show owners, managers of public and private menageries, and even foreign explorers. Eventually, his collection of carefully preserved specimens included ocelots, antelopes, monkeys, seals, bears, hyenas, leopards, lions, tigers, and the jaws of at least two elephants. He even managed to get hold of a whale carcass that had washed up along the Thames, which he lovingly disassembled.

As Hunter's anatomical skills grew, so too did his reputation among the ranks of England's naturalists. They started seeking his opinion and requesting his assistance with particularly unusual or difficult specimens. The eminent naturalist John Ellis received a shipment of odd amphibians from a colleague in South Carolina. They were about two feet long, had a head and front legs similar to a salamander, with feathery external gills and the bodies of eels. Ellis sent one to the renowned Swedish taxonomist Carl Linnaeus, who proclaimed it an entirely new genus and species. He used his naming system to give it the scientific name *Siren* (Latin for "mermaid") *lacertina* (little lizard).

Ellis retained the other specimens and asked Hunter for help investigating their internal structure. Hunter discovered that the creatures not only had a set of external gills but also an internal pair of lungs. He declared it a missing link between fish and amphibians. Ellis and Hunter reported their results to the Royal Society in June 1766. Eight months later, John Hunter was elected a fellow of the Royal Society and recognized, "as a person well skilled in Natural History and Anatomy." The recognition must have been made all the sweeter by the fact that John had beaten William's election to the society by three months.

Hunter's burgeoning collection, now including a siren's heart preserved in a jar, began taking up every bit of spare room in his house. To relieve the overcrowding, in 1765, he bought a parcel of farmland in Earls Court, outside of London, and had a nice two-story house built to accommodate his scientific pursuits.

With the luxury of this additional space, Hunter was able to add a collection of live animals to his preserved specimens. Visitors to the home were greeted by all manner of howls, barks, bleats, and snarls as they approached the large iron gates. Once on the grounds, they would have seen a fishpond decorated with animal skulls and the front doors of the house adorned with crocodile jaws. On the lawn grazed zebras, mountain goats, and Asiatic buffalo alongside the more typical sheep and cattle. From the kennel could be heard various dogs and jackals, including Hunter's favorite, a large wolf-dog hybrid given to him by an animal dealer named Wild Beast Brooks. A large hillock in the garden housed lions and leopards, and around the buildings was a six-foot-deep trench that ended in a sturdy set of wooden doors, leading to Hunter's underground laboratory.

Hunter was kept busy tending to the needs of his growing menagerie but found time to treat animals belonging to his neighbors and patients. This led him in 1791 to help launch the Veterinary College of London, forerunner to the Royal Veterinary College. His exotic pets, his interest in veterinary science, and undoubtedly the frequent sight of him driving his carriage into town, pulled by three of his Asiatic buffalo, caused many to speculate that John Hunter may have been the inspiration for Hugh Lofting's eccentric physician, who talked to animals, Dr. John Dolittle (Magyar, 1994). An additional tidbit adding credence to the theory is a well-known letter from Hunter responding to a request for advice on a case from one of his students.

Passing along the lessons he learned on Belle Île, he wrote, "I believe the best thing you can do is to do little" (Rains, 1976).

Hunter's influence, however, stretched far beyond inspiring colorful literary characters. In December 1768, he was elected a staff surgeon at St. George's Hospital. The position offered no salary but did give Hunter the opportunity to begin teaching. Like Cheselden before him, Hunter instilled in his students a deep and abiding respect for applying the tools of science to their medical craft and an abhorrence of blindly repeating the mistakes of the past. For the next twenty-five years, the gruff but charismatic Scotsman taught his lessons well to over a thousand medical students. Among them was one of his first and perhaps favorite students, Edward Jenner, who would go on to kindle his own medical revolution when he applied Hunter's methods to the development of the first vaccine.

Many of Hunter's students went on to become great surgeons and teachers in their own right, spreading their teacher's scientific approach to their own students. Others of Hunter's disciples carried his methods across the Atlantic to America. William Shippen and John Morgan, in particular, used Hunter's teachings as the founding philosophy for America's first medical school, the College of Philadelphia, which later became the University of Pennsylvania.

By the middle of the 1770s, Hunter's life showed all the signs of success. He and Anne were happily married, with a growing family. His medical practice was flourishing. He had the admiration of his students and the respect of his scientific peers. It seemed as if only one victory eluded him, victory over death itself. Hunter's third child, James, died while still an infant. He was followed in death the following year by his older sister, Mary-Anne. Even though child deaths were common in the eighteenth century and despite the comfort of his two remaining children, John and Agnes, it must have been a bitter loss for the mighty surgeon.

Concerned about the alarming number of drowning deaths, a local apothecary named William Hawes, along with the physician Thomas Cogan, resolved to do something about it. In 1774, they formed the Humane Society, later becoming the Royal Humane Society, and approached the famous surgeon John Hunter for help. To promote their cause, the group offered a reward of four guineas to anyone who succeeded in reviving any person "taken out of the water for dead" within thirty miles of London.

Hunter provided his own views on resuscitation for the group and later presented them to the Royal Society. He argued that drowning victims should not automatically be considered dead. Rather, he suggested "that only a suspension of the actions of life has taken place." He went on to suggest that promptly taking certain steps could restore them to life. The first should be to return air to the victim's lungs. Hunter suggested using a double bellows to pump air into the person's nose and mouth. He even suggested Priestley's recently discovered dephlogisticated air could be used for this purpose.

The second step was to stimulate the victim's heart. Hunter suggested holding stimulating vapors under the person's nose, warming the victim and rubbing the body with essential oils. If that failed, he proposed using electrical shocks to restart the heart. He wrote, "Electricity has been known to be of service, and should be tried when other methods have failed." Above all, he advocated the scientific approach and recommended that an accurate journal be kept with descriptions of the methods used and their degree of success.

Hunter got a chance to try out his radical resuscitation techniques on June 27, 1777, not on a hapless drowning victim, but on a convicted criminal. The Reverend William Dodd was a popular and flamboyant figure in Georgian London. Although a cleric, he seemed to devote as much time to good living as to the Good Book. His extravagant lifestyle and fine dress earned him the nickname the "Macaroni Parson." Unfortunately, his lifestyle got the better of him, and to meet ever rising debts, he forged a bond for forty-two hundred pounds. The amateurish attempt failed, and he was convicted and sentenced to hang.

Despite the conviction, Dodd remained a popular figure. He had made substantial contributions to charities such as the Magdalen House for "fallen women" and the recently formed Humane Society. The famous writer Samuel Johnson argued his cause, and a petition for leniency with twenty-three thousand signatures was presented to George III, all to no avail. His only hope was in John Hunter's ability to cheat death.

Hunter knew from his years dissecting executed criminals that executioners were often careless about their craft. Rather than dropping the convicts from a sufficient height to quickly snap their necks, they frequently allowed only enough rope to allow their victims to dangle and strangle to death as their oxygen supply was slowly cut off. This led to a few well-publicized cases of executed criminals surviving their executions, only to revive some time later on the dissection table. This gave Hunter the perfect chance to test his theories, so while Dodd ascended the gallows, Hunter and his Humane Society cohorts waited at a nearby undertaker's, with bellows and essential oils at the ready.

Hunter never wrote of what happened next, but one of his friends, fellow Royal Society member Charles Hutton, recounted the story that Hunter told him in the privacy of a local coffeehouse. Hunter had hoped to get hold of Dodd's body as quickly as possible, but the wayward cleric remained hanging for almost an hour before being cut down. It took an additional forty minutes for the hearse to deliver him into Hunter's hands. Undeterred, the surgeon and his associates "tried all the means in their power for the reanimation." Alas, they were unsuccessful, and Hunter was forced to admit that what might have been his greatest experiment was a failure. In spite of that, for the next several years, the press reported a series of Elvis-like sightings in and around Great Britain of Dodd alive and well. As late as 1794, a Scottish newspaper reported that he was living in Glasgow.

While the press was busy chasing down Dodd sightings, Hunter was moving on. Even with his Earls Court estate and the house he retained in the city, Hunter was running out of room for his ever-growing collection. It wasn't simply a hodgepodge of random curiosities. It was a carefully organized series of exhibits with which Hunter wished to illustrate what he had learned about the miracle of life. It was his legacy, and he needed a proper place to share it with the public.

He found a large house in Leicester Fields, later known as Leicester Square. It was an elegant four-story home, with a large entry hall, a study and a parlor for Hunter to see patients. Upstairs was a large drawing room, perfect for Anne to host parties with her society and literary friends. Above that, were two more floors, and beneath it all was an underground stable. As large as it was, it was still insufficient for Hunter's needs, so he purchased the house in back that fronted on Castle Street, later renamed Charing Cross Road, and the land in between. He then hired a veritable army of workmen to connect the two properties.

It took two years to complete, but when it was finished, the custom-built glass-and-brick building boasted a lecture hall, a grand parlor, and Hunter's private museum. Stunned visitors beheld one of the greatest natural history collections of the eighteenth century, featuring thousands of Hunter's medical specimens, all manner of stuffed and mounted animals, including a full-size giraffe and the jaw of a full-grown bottle-nosed whale. Unseen by the guest was Hunter's dissecting room in the attic, his private printing press, and quarters for his many students and assistants. The back of the house, which had formerly been the front of the Castle Street house opened onto a busy thoroughfare. Next to it was a ramp leading down to the subterranean stables where Hunter could have his less savory specimens discreetly delivered.

So impressive was the structure, and such a well-known landmark did it become, that a hundred years later it would inspire Robert Louis Stevenson. While he was writing *The Strange Case of Dr. Jekyll and Mr. Hyde* in 1886, he used John Hunter's former home as the model for the home of his protagonist, Dr. Henry Jekyll (Moore, 2005). Stevenson writes that Dr. Jekyll had purchased the home from "the heirs of a celebrated surgeon." He describes the front rooms where the good doctor sees his patients and the lecture hall "once crowded with eager students." The murderous Mr. Hyde enters and exits through the rear of the house, which opens onto a dingy thoroughfare.

Over the next several years, the good John Hunter entertained numerous prominent guests and counted among his patients some of the best-known figures in England. When his friend Benjamin Franklin was suffering from bladder stones, Hunter was consulted. When the Scottish economist Adam Smith required surgery for hemorrhoids, he called in his fellow Scotsman to perform the operation. Smith was so pleased with the outcome that he sent Hunter a copy of his great work *The Wealth of Nations* and recommended the surgeon for two army appointments, stating, "nothing is too good for our friend John" (Smith, 1787).

In 1788, when George Gordon Byron, later Lord Byron was born, it was Hunter who inoculated the infant against smallpox and recommended that a special boot be used to correct the boy's twisted foot. Unfortunately for Byron, his mother was unable or unwilling to follow Hunter's advice, and the future poet's clubfoot troubled him for the remainder of his life.

Although he was lauded by his patients and idolized by his students, there were also many who resented Hunter. Many of his fellow surgeons and physicians at St. George Hospital were angered by his unorthodox methods and jealous of the students who flocked around him. They opposed him at every opportunity.

When Hunter was appointed surgeon general of the army, it forced him to divide his attention. His opponents took advantage, passing a number of onerous duties and responsibilities on surgeons in an attempt to make him quit. When that failed, they tried to place restrictions on incoming students to limit his influence. In one final battle on October 16, 1793, Hunter confronted his critics over the admission of two young Scottish surgical students. While vociferously arguing their case, Hunter, who had a history of angina, collapsed and died. The autopsy, performed on

Hunter's own dissecting table, confirmed the cause of death as his longstanding heart condition, likely aggravated by the syphilis he contracted as part of his own experiments.

By any standard, John Hunter was a remarkable figure. He dragged surgery, kicking and screaming, from its medieval traditions, and applied the scientific rigor by which surgeons practice today. He angered and amazed, frightened and fascinated the British public until the end of the eighteenth century, and left behind myriad beautifully preserved specimens still capable of inspiring wonder in visitors to the museum that bears his name. If such an incredible life and philosophy could be briefly summed up, it's best done by Hunter's own words. In a 1775 letter to his beloved student Edward Jenner, who frequently asked for his mentor's opinion, Hunter wrote, "I think your solution is just, but why think? Why not try the experiment?"

As fierce as the battles Hunter endured, they were nothing compared to the medical war waged by one of his later comrades in Vienna. There, in the hallowed halls of the Allgemeine Krankenhaus, The General Hospital of Vienna, one of the most respected teaching hospitals in all Europe, a lonely and misunderstood crusader would launch a campaign against medical intransigence, political in-fighting and blatant denial as pernicious as the disease it allowed to spread. At stake were the lives of thousands of mothers and children. Defending them was an irascible and steadfast Hungarian named Ignác Semmelweis.

Straddling the banks of the Danube River sat the Hungarian cities of Buda on one side and Pest on the other, both then part of the Habsburg Empire. On the Pest side, on July 1, 1818, the rising bourgeois family of József and Terézia Semmelweis welcomed their fifth child, Ignác. Ignác's father was a prosperous grocer and his mother came from a wealthy Bavarian family, so, Ignác was the beneficiary of a comfortable middle-class upbringing. Although born in Hungary, he grew up speaking the Germanic Buda-Swabian dialect of his father and the Bavarian German of his mother. He wouldn't properly learn Hungarian until he entered secondary school and thus retained his unique accent for the rest of his life.

Ignác, or Naci as he was called by his family, did well, although not exceptionally so, in school, and in 1837 he entered the University of Vienna. He intended to follow his father's wishes and study law, but while attending an anatomy lecture with a friend, he heard the charismatic professor Josef Berres. So fascinated was the young Semmelweis, that the following year, he enrolled in the university's medical school.

He studied medicine at the University of Vienna for a year, then transferred to the University of Pest for two years, and then back to Vienna, where he graduated. It seems like a circuitous route, but it was a common one. The University of Vienna was considered a better school, and students who graduated from there could practice anywhere in the Austrian Empire. Graduates from the University of Pest were restricted to practicing in Hungary.

The early 1840s were a turbulent time to be a Hungarian university student. Hungary had long been subject to Austrian rule, and treated as one of Austria's poor relations. Hungarian dissatisfaction took root, and the universities provided fertile ground for it to grow. All across campus and in nearby coffeehouses, disaffected students expressed their anger of stifling government and university policies. Semmelweis found himself immersed in this heady environment of radicals. Eventually this burst forth into the uprisings of 1848, under the banner, "Freedom of teaching and freedom of learning."

Nowhere was the need for reform more acutely felt than in the university's medical school. Hidebound and under the thumb of government ministries, the faculty defended the status quo. The German surgeon Theodor Billroth later described the University of Vienna professors as:

> [a] generation that had been reared in an intellectual straitjacket with dark spectacles before their eyes and cotton wool in their ears. The young people turned summersaults in the grass, and the old men, whose bodies had been hindered in their natural development by the lifelong burden of state supervision, felt their world tumbling about their ears, and believed that the end of things was at hand (Nuland, 2003).

By February 1844, Semmelweis had finished his dissertation and was ready to graduate. Just as he was preparing for the spring graduation and looking forward to his medical career, he received the shattering news that his mother had unexpectedly died. He rushed back to Buda to be with his family, pausing only to sign a statement in the faculty register affirming his intention to return.

Within six weeks he was back, pursuing his career and applying for assistantships. Thinking back to the thrilling anatomical lecture of Berres, he first applied for a position with the up-and-coming anatomist Jakob Kolletschka, who was making significant contributions to the emerging field of forensic pathology.

Semmelweis hoped to make his own contributions and anxiously applied to assist Kolletschka. Unfortunately, the competition was stiff and his application was denied.

Undeterred, Semmelweis applied with the school's leading physician, Joseph Škoda. He was using the newly fashionable tools of percussion and what was called auscultation. Long before the stethoscope became a common sight around the necks of doctors, Škoda was using the sounds of the body to diagnose disease. Again determined to make a name for himself, Semmelweis applied for a position. Again the position went to someone else.

Having struck out twice, Semmelweis began to set his sights a bit lower. Obstetrics wasn't considered an especially prestigious field in the mid-nineteenth century. The course was an elective, and most deliveries at the time were carried out by midwives. What's more, the obstetrics department at the General Hospital of Vienna was under the direction of a conservative older physician named Johann Klein. He was a political appointee who dedicated himself to sticking to the curriculum and not rocking the boat. Klein accepted the application of the young Semmelweis, completely unaware of what he was getting himself into.

From the moment Semmelweis set foot on the obstetrics ward, it was apparent that a crisis was taking place. The lying-in unit, as it was called, was divided into two halves, the First Division, staffed by doctors and medical students, and the Second Division, staffed by midwives. Most women who could afford to were still delivering at home, so many of the patients in both divisions where poor women with no place else to go. As Semmelweis walked among the beds of indigent women in the First

Division, he was horrified to learn that one in every six of them was dying of a mysterious disease called puerperal fever or, simply, childbed fever.

The symptoms were unmistakable. Within twenty-four hours of delivering, the mother would begin running a high fever. Her abdomen would become sore and inflamed, and she would suffer nausea and headaches accompanied by a foul-smelling vaginal discharge. Over the course of the next day or two, she would alternate between coma and fevered delirium. Almost invariably these patients died. The doctors and nurses could do little to help them.

Such tragic death wasn't confined to Vienna. In obstetrics wards all across Europe and America, similar waves of maternal death were happening. In the words of Charles Delucena Meigs, one of the period's most respected obstetricians in the United States:

> There is a "word of fear" that I shall pronounce when I utter the name of Puerperal Fever; for there is almost no acute disease that is more terrible than this. . . . There is something so touching in the death of a woman who has recently given birth to her child; something so mournful in the disappointment of cherished hopes; something so pitiful in the deserted condition of the newborn helpless creature, for ever deprived of those tender cares and caresses that are necessary for it— that the hardest heart is sensible to the catastrophe. It is a sort of desecration (Nuland, 2003).

This wasn't a new phenomenon. Puerperal fever had been known for centuries. Hippocrates wrote of such deaths and well-known historical cases include Jane Seymour, third wife of Henry VIII, and Mary Wollstonecraft, who died after giving birth to her daughter, the future Mary Shelley. What was so alarming was that since the mid-eighteenth century, what had once been a relatively rare occurrence was now becoming an epidemic.

Desperate to find the cause, doctors performed autopsies, but instead of answers, they found only the bodies of formerly healthy young women, the very source of life itself, now lying upon the morgue tables, their abdomens distended and filled with pockets of foul-smelling pus. Even more baffling to Semmelweis and his fellow physicians was that the Second Division, staffed by midwives, had no such epidemic. There would be sporadic cases, but only a small fraction of the number of cases in the First Division. This disparity was evident to everyone, including the patients. Semmelweis recounted:

> That they were afraid of the First Division there was abundant evidence. Many heart-rending scenes occurred when patients found out that they had entered the First Division by mistake. They knelt down, wrung their hands and begged that they might be discharged. Lying-in patients with uncountable pulse, meteoric abdomen [rapid onset of distention], and dry tongue, only a few hours before their death, would protest that they were really quite well, in order to avoid medical treatment, for they believed that the doctor's interference was always the precursor of death (Nuland, 2003).

In the midst of this terror, the doctors were baffled. Various causes of puerperal fever had been postulated, ranging from too tight corsets to poor diet to drafts in the delivery room or any unspecified shock to the mother's system. In 1789, a Scottish physician named Alexander Gordon even proposed the outlandish theory that it was caused by some contagious agent being carried to the patients by the doctors and nurses themselves. He was ignored. The American writer/physician/jurist Oliver Wendell Holmes advocated similar views, with similar effect, or lack thereof.

The most widely held view was that puerperal fever was caused by some sort of miasma or bad air. The windows of delivery rooms were ordered opened and some obstetrics wards went so far as to drill holes in the window sashes and doors to increase ventilation. All to no avail, the epidemic continued.

Perhaps because of the recent death of his mother, Semmelweis threw himself into the search for the disease killing so many other mothers. Every morning he went to the morgue to perform autopsies on the recently deceased women. That time of day, the morgue or "dead house" as it was known, was busy with other medical students conducting autopsies or learning how to perform gynecological exams on the bodies of deceased women, a common practice of the time.

Semmelweis meticulously recorded his findings from the autopsies and correlated them against hospital records. Slowly a pattern began to emerge. Both divisions had similar numbers of patients, between thirty and thirty-five hundred per year. Deliveries in the First Division were performed by doctors and medical students, and in the Second Division, by midwives. Of those cases, the First Division experienced six to eight hundred maternal deaths due to puerperal fever, or between 25 and 30 percent. In the same period, the Second Division could expect approximately sixty such deaths, less than 10 percent of the other division. There also seemed to be a correlation between the amount of trauma experienced by the mother during delivery and her chances of dying. Finally, when a mother died of puerperal fever, her newborn child would frequently die as well. Semmelweis conducted autopsies on these babies, and found that their bodies mirrored the swollen and pus-filled conditions of their mothers.

Semmelweis also noticed that if First Division were temporarily shut down, the incidence of puerperal fever dropped dramatically. When the unit was reopened, the incidence would return to its previous rate. Klein, the head of the unit blamed the poor conditions of the walls for the outbreaks. When he presented this at a meeting in 1846, Semmelweis, still a junior physician, rose and pointed out that the walls in other maternity hospitals were frequently in worse shape, but their mortality rates were lower. Klein didn't take being contradicted by one of his subordinates lightly. It caused a growing animosity between the two men.

The final clue to the cause of the deaths came at the cost of a promising doctor's life. Since he first applied to assist Kolletschka, Semmelweis and the young forensic pathologist had become friends as well as colleagues. On March 20, 1847, while Semmelweis was on a brief vacation, Kolletschka was accidentally cut on the finger by a student during a routine autopsy. Shortly after that, the wound became infected, he developed a high fever, and died a few days later. When an autopsy was performed, Kolletschka's body displayed the same signs of massive, systemwide infection that appeared in the bodies of childbed fever victims. Semmelweis looked at the evidence of his friend's death:

Totally shattered, I brooded over the case with intense emotion until suddenly a thought crossed my mind; at once it became clear to me that childbed fever, the fatal sickness of the newborn and the disease of Professor Kolletschka were one and the same, because they all consist pathologically of the same anatomic changes. If, therefore, in the case of Professor Kolletschka general sepsis [contamination of the blood] arose from the inoculation of cadaver particles, then puerperal fever must originate from the same source. Now it was only necessary to decide from where and by what means the putrid cadaver particles were introduced into the delivery cases. The fact of the matter is that the transmitting source of those cadaver particles was to be found in the hands of the students and attending physicians (Nuland, 2003).

Every morning, the medical students dutifully performed autopsies, often on the bodies of mothers who had succumbed to childbed fever. They then went directly to the ward where they gave internal exams to their obstetric patients. The advent of germ theory of disease was still some time off, so they saw no need to wash their hands. At best, they might wipe the blood, pus and other bodily fluids of the deceased off on their coats or pants, or given their hands a cursory rinse in cold water. Many of them even had a certain amount of pride in the "hospital smell" their hands carried, wearing it as a sort of badge of honor of a true physician.

Semmelweis conducted a bold, if obvious, experiment. That spring, he instituted a rule requiring all personnel entering the First Division first wash their hands thoroughly with soap and water. Basins and stiff nailbrushes were provided at all the entrances. Eventually, the soap and water were augmented with a chlorine bleach solution. The staff was initially resistant to go through what they saw as an arduous and useless procedure, but when Semmelweis refused to compromise, they reluctantly complied.

The results were rapid and dramatic. Within a month, the First Division's death rate due to childbed fever plunged from the typical 25 percent to below 2 percent, a rate closely matching that of Second Division. After a year of enforcing Semmelweis' hand washing procedures, the death rate went down closer to 1.2 percent. Many at the hospital, particularly younger physicians, grasped the significance, and joined Semmelweis' cleanliness campaign. Among his converts were Rokitansky and Škoda. Klein, however, and the rest of the old guard remained unconvinced. They questioned how a young rebel, a Hungarian, no less, could bring into question the methods that they had been using for years.

Semmelweis' case wasn't helped by the fact that he resisted publishing his results. It wasn't that he lacked the courage of his convictions. More likely it was due to his lack of confidence in his own writing ability. German was the lingua franca of the medical community, and Semmelweis, a Hungarian of Buda-Swabian heritage, was never quite fluent in that language. Despite his failure to publish, word of Semmelweis' discovery got out. One of his friends, Ferdinand von Hebra, publisher of a prominent Viennese medical journal, printed brief notices of Semmelweis' findings in December 1847 and April 1848. Semmelweis and his followers also circulated a series of open letters to prominent medical practitioners inviting comments.

Semmelweis' critics, however, were not hesitant to publish. The conservatives mercilessly attacked the young upstart and anyone who supported him. Medical journals were filled with the most virulent attacks on the new theory. Klein and his allies argued that it was impossible that they, the most highly trained and most esteemed medical professionals in all Europe could be the cause of their patient's deaths.

The reaction was understandable. Admitting that Semmelweis was right would mean admitting their own guilt. Semmelweis himself felt the weight of this burden:

Because of my convictions, I must here confess that God only knows the number of patients who have gone to their graves prematurely by my fault. I have handled cadavers extensively, more than most accoucheurs [male obstetricians or midwives]. If I say the same of another physician, it is only to bring to light a truth, which was unknown for many centuries with direful results for the human race. As painful and depressing, indeed, as such an acknowledgment is, still the remedy does not lie in concealment and this misfortune should not persist forever, for the truth must be made known to all concerned (Nuland, 2003).

Semmelweis was not alone in his admission of guilt. Word of his discovery found its way to Professor Gustav Adolf Michaelis, a prominent teacher of obstetrics in Kiel, Germany. Michaelis decided to test Semmelweis' hypothesis by instituting similar hand-washing procedures. His results confirmed Semmelweis' findings. However, shortly before obtaining this incontrovertible evidence for the cause of childbed fever, he had attended at the birth of his niece. The child died a few days later from puerperal fever. Faced with the crushing realization that he had caused the death of his beloved niece, Meichaelis committed suicide by throwing himself beneath an oncoming train.

In March 1849, Semmelweis' position came up for renewal. Klein seized the opportunity and refused to renew it, citing what he called the younger man's autocratic behavior and the tensions it created at the clinic. Semmelweis responded by applying for the position *Privatdozent* of midwifery, a private physician with teaching privileges at the school.

It should have been an easy appointment, but for seventeen months Semmelweis waited for word whether his application was accepted or denied. When it finally was approved, it came with restrictions attached. Semmelweis was forbidden from using cadavers to teach with and was forced to demonstrate procedures on a mannequin. He was also unable to issue certificates of attendance to his students, a privilege the other instructors enjoyed.

In the face of these galling restrictions, under continual attack from the establishment and suffering from his own guilt, Semmelweis fled Vienna. He packed up his bags and left for home without even telling his closest friends. Many of his allies, including Rokitansky, Škoda, and Hebra, were left to feel that he had abandoned the fight.

The Buda-Pest that Semmelweis had left as an optimistic young student was a far cry from the one he returned to. His father died while he was away. The uprisings of 1848 had been brutally put down and the authorities instituted rigid controls to prevent further disorder. By the time Semmelweis returned, three of his brothers were living in exile.

After months of looking for a place to work, in May 1851

Semmelweis managed to secure an unpaid position in the obstetric division of St. Rochas Hospital in Pest. This hospital too was suffering an epidemic of childbed fever, and shortly after Semmelweis arrived, he managed to convince the administration to initiate his hand-washing regimen. Again the staff was resistant. Again Semmelweis refused to compromise. As expected, the death rate plummeted.

Physicians at the hospital were impressed and over the next several years, Semmelweis established a successful practice. By 1855, he had been appointed professor of theoretical and practical midwifery at the University of Pest. As his prospects improved, he even found time to fall in love with a lovely woman half his age. He married the twenty-year-old Maria Wiedenhoffer, the daughter of another Buda-Swabian merchant in 1856.

Emboldened by his success, Semmelweis finally found the confidence to write about puerperal fever. In 1858, he published "The Etiology of Childbed Fevers" in a Hungarian journal. It met with predictable attacks by the conservative medical establishment. What's worse, since Semmelweis' departure from Vienna, his protocols at the General Hospital had been abandoned, and the death rate once again soared. In a desperate attempt to silence his critics, Semmelweis resolved to write a full account of his findings in German so it might be presented to the medical community as a whole.

He published his magnum opus, *The Etiology, the Concept and the Prophylaxis of Childbed Fever* in 1860–1861. It did his cause more harm than good. The first half of the book was an exhaustive, overly dense description of the disease and his experiences with it in Vienna. Much of it was repetitive and poorly organized, filled with redundant graphs and tables. It was, at best, a difficult read.

The second half was an attempt to address the specific claims of his critics, but in his zeal, Semmelweis completely ignored the common rules of engagement of scientific debates. Instead, he gave full vent to his spleen and unleashed on his critics a barrage of accusations and name calling. He refers to one of his critics, Friedrich Wilhelm Scanzoni, a professor at the University of Prague, as a "wretched observer." In addressing one of Germany's most eminent pathologists, Rudolf Ludwig Karl Virchow, who would later be considered one of the fathers of cell theory, Semmelweis says, "823 of my students are now midwives practicing in Hungary. . . . They are more enlightened than the members of the Society of Obstetrics in Berlin; they would laugh Virchow to scorn if he attempted to lecture them on epidemic puerperal fever" (Carter, 1985).

What Semmelweis intended to be his masterful counterattack now became further fuel for his critics. They savaged him in the professional journals and many did all in their power to suppress his book. So incendiary was Semmelweis' rhetoric that even his supporters drew back from the heat. Others simply turned their backs and tried to remain above the fray. In response, Semmelweis only intensified his attacks. He sent out another series of open letters addressed to his opponents. In one he wrote to Joseph Spath, professor of obstetrics at the University of Vienna:

Within myself, I bear the knowledge that since the year 1847 thousands and thousands of puerperal women and infants who have died would not have died had I not kept silent, instead of providing the necessary correction to every error which has been spread about puerperal fever. . . . And you, Herr Professor, have been partner in this massacre. The murder must cease, and in order that the murder ceases, I will keep watch, and anyone who dares to propagate dangerous errors about childbed fever will find in me an eager adversary (Nuland, 2003).

Semmelweis' condemnation of Scanzoni was full of sarcasm and equally dire:

Should you, however, Herr Hofrath [a term used to address respected professors], without having disproved my doctrine, continue to train your pupils in the doctrine of epidemic childbed fever, I declare before God and the world that you are a murderer and the "History of Childbed Fever" would not be unjust to you if it memorialized you as a medical Nero, in payment for having been the first to set himself against my life-saving theory (Nuland, 2003).

Perhaps the years of fighting, of seeing the carnage wrought by medical intransigence and outright denial had taken too high a toll. There are many who mark this as the beginning of Semmelweis' descent into madness. His health began to noticeably deteriorate, and the man, who as a young physician had been described as "lighthearted" and "popular" with a "playful and jocular nature," became increasingly combative and isolated. He seemed to alternate between periods of depression and apparent hyperactivity. He had difficulty sleeping, and would frequently roam the house, sometimes the streets, talking to himself.

His wife, Maria, attempted to care for him as best she could, but when it became apparent that his mental state was worsening, his friends convinced her that more drastic steps were necessary. In July 1865, she convinced her husband that he should go to a spa in Grafenburg to recover his health. In reality, she was preparing the way for him to enter a mental hospital. On July 29, Maria accompanied by her uncle and Semmelweis' assistant helped her husband onto the overnight train to Vienna. When they arrived, Semmelweis was met on the platform by his old friend von Hebra, who convinced him to come see the new sanitarium he had opened. They took him instead to the public insane asylum.

While Semmelweis was distracted talking to one of the staff, Maria and von Hebra signed the commitment papers and slipped out. Years later, Maria Semmelweis recounted what happened the following day when she returned to check on her husband: "Hofrath Riedel, director of the sanatorium, met us in person to say that the night before my husband had tried to get out and, when he was forcibly restrained, fell into a fit of delirium so that six attendants could scarcely hold him back. I was not allowed to see him." Two weeks later, she received word that he was dead.

Semmelweis' body was sent to his former hospital. An autopsy was performed on the very table on which his friend Kolletschka had been examined. Like his friend, Semmelweis' body showed all the signs of a systemwide septic infection. In other words, he died from puerperal fever, a victim of the disease he had spent his life fighting. The initial story was that Semmelweis had injured his finger in much the same way Kolletschka had while performing an internal exam on a patient.

Almost a century later, in 1963, Semmelweis' body was exhumed, and an investigation was undertaken revealing the truth. Ignác Semmelweis had received a brutal beating at the hands of the asylum guards and was left without medical attention. The wounds became infected and he likely died in the same type of fevered delirium that he had so often seen in his patients.

Years after his death, Semmelweis would be vindicated by the work of fellow pioneers Louis Pasteur and Joseph Lister as they proved the link between germs and disease and demonstrated the utility of antiseptic techniques. Ignác Semmelweis was by nationality, politics, and disposition an outsider, who was forced into the unlikely roll of medical crusader. His uncompromising stance would enrage his opponents and often alienate his allies. In the end, his battle cost him his career, his sanity, and ultimately his life, but millions of women and their children owe their lives to his heroic struggle.

After the investigation was finished, Semmelweis' remains were reinterred in the courtyard of the house where he was born, now a medical museum named after him. Hungary honored its native son with a statue in Budapest. It features Semmelweis standing with a book in his hand; at his feet is a woman with a baby in her arms looking up at the man who saved her life.

CHAPTER FIVE

Electrical Spectacles
and the Spark of Genius

AROUND THE TIME THAT JOHN HUNTER WAS DELVING
into the macabre mysteries of the human body in the 1700s,
there was another young anatomist making his own contributions to the legend of the mad scientist. There were, of course, a
number of noticeable differences between him and Hunter. Unlike the famous surgeon's Scottish roots and modest upbringing,
he was an Italian, from a professional family. He wasn't skulking
in churchyards and relying on the tender mercies of the grave
robber for his materials. Rather, he was a sheltered academic,
working from the confines of one of Europe's most prestigious
universities. While Hunter was methodically searching for new
insights, he would make his most famous discovery quite by accident, but that accident would lead to shocking discoveries. In
fact, his name would one day become closely associated with the
phenomenon of electricity. That name was Luigi Galvani.

Luigi Galvani was born on September 9, 1737, in the city of Bologna in northern Italy. Not much more than that is known of his early life. What is known is that he was the son of a physician, and his early ambition was to study theology and enter a monastery. However, he was eventually persuaded by his family to follow in his father's footsteps. He went on to study medicine at the University of Bologna, one of the oldest and most respected universities in Europe.

Galvani undertook his medical studies there, and showed a particular interest in anatomy and physiology. He wrote his thesis on the nature and formation of bones. In 1762, at the age of twenty-five, he was awarded degrees in both medicine and philosophy. Not long after that, he was made a lecturer of anatomy at the university. It is interesting to note that, despite his successful career in medicine, Galvani never lost his religious passion. He never ended a lecture without reminding his listeners of, "that eternal Providence, which develops, conserves, and circulates life among so many diverse beings" (*The Catholic Encyclopedia,* 1907).

Galvani did most of his research on and produced a number of significant writings on comparative anatomy. One of his best regarded early works was a treatise on the kidneys of birds. He also did quite a bit of research into the sensory systems of various animals, which led him to an interest in the nervous system.

Coincidentally, around this time, Joseph Priestley's book *The History and Present State of Electricity* was translated into Italian. Galvani was fascinated by Priestley's descriptions of electrical phenomena and his accounts of Franklin's exploits. In hopes of duplicating some of them himself, he obtained Leyden jars and a spark generator, similar to the one Priestley had used, and began conducting his own electrical experiments. However, he didn't neglect his anatomical studies. He simply conducted his research in both fields at the same time and in the same laboratory.

It was in the midst of this cross-disciplinary study that his fortuitous accident happened. Galvani stumbled on what would become one of electricity's greatest discoveries. Around 1780—the exact date is in some dispute—while performing anatomical research, he was in the process of dissecting a frog. At the same time, one of his assistants was working with a hand-cranked machine for generating electric sparks. As the assistant reached for one of the metal dissection knives near the skinned and partially dissected frog, an electric spark jumped from the knife to the sciatic nerve in the frog's leg. The leg immediately convulsed, giving the dead frog the appearance of life.

Startled by this, Galvani tried poking the leg again with a scalpel to ensure that it was dead. Since the generating machine was no longer being cranked, the leg simply lay there unmoving. This started him to wondering, could there be a link between the contraction of the leg and the sparks from the machine? Could electricity somehow animate the unliving tissue of the frog, and what was the nature and source of the electricity?

Galvani immediately began to investigate the possibility. Being a trained scientist and veteran experimenter, he carefully set up a series of tests to replicate the phenomenon. In his own words, "For it is easy in experimentation to be deceived, and to think one has seen and discovered what we desire to see and discover."

Galvani tried to see if the material that he touched the frog leg with had an effect. An iron rod duplicated the original results, but a glass rod did not. He tried attaching the leg to a long wire while he cranked the generator. When he touched the opposite end of the wire, the leg again convulsed. Again and again he repeated his experiments, changing one variable and then another.

In one of his more ghoulish attempts, he attached the spinal cords and legs of several freshly dissected frogs to brass hooks. Then he took them to a garden terrace at the Palazzo Zamboni near his home. He hung the hooks from an iron railing that surrounded the garden and waited for an approaching storm. Soon he was rewarded by the gathering clouds and distant flashes of lightning. As the electrical storm drew nearer and nearer, the frog legs, dangling from their hooks, began to jerk and twitch. Imagine the sight of Galvani, sitting there in the garden watching in delight as his disembodied frogs' legs danced to the accompaniment of thunder and lightning.

Still, Galvani wasn't satisfied. He repeated the same experiment, hanging the frog legs from brass hooks and then suspending them from an iron railing, only this time, he did it on a sunny day, with no storms nearby. To his amazement, the legs again began to dance. At first, he had thought that the storm was the source of the electricity needed to produce the effect. Now he became convinced that somehow the frog legs themselves were the source.

Over the course of ten years, Galvani used these and subsequent experiments to develop his theory of animal electricity, what today we would call bioelectricity. In 1791, he published his results in *De Viribus Electricitatis in Motu Musculari Commentarius* (Commentary on the Effects of Electricity on the Motion of Muscles). It was met with widespread acclaim. One of his colleagues, Alessandro Volta, a professor of physics at the University of Pavia, and himself acutely interested in electricity, declared Galvani's theory, "among the demonstrated truths." Volta even named the phenomenon Galvanism, after its discoverer.

The story might have ended there, but Volta became so interested in Galvani's discoveries that he decided to reproduce the experiments himself. Galvani, being an anatomist, was predisposed to look to the body of the frogs as the source of the electricity. Volta, on the other hand, was a physicist. He conducted his own experiments and showed that the frog's muscles would contract even when they were not part of an electrical circuit. He could produce the same effects simply by touching the leg with two probes made of different types of metal. Volta soon became convinced that the phenomenon was produced by the metal itself, and when he published his results, instead of referring to it as "animal electricity," he renamed it metallic electricity.

What followed was one of the great disputes of scientific history. Galvani and his supporters countered Volta's claims by showing that they could produce the twitching in the legs using two identical pieces of metal. Not so fast, responded the Volta camp. Even if the two pieces of metal looked the same, there could be small differences that could account for the results. Back and forth it continued.

Eventually, Volta showed that he could produce a similar effect without any frog legs at all. He used alternating plates of different metals, copper and zinc, and instead of an amphibian's appendage, he separated them with pieces of cardboard soaked in saltwater. With this primitive apparatus, he was able to produce electricity. He had created his previously mentioned voltaic pile. In reality, it was the first battery. Still the battle raged on.

There are two interesting things to note about the controversy. First, unlike many of their supporters, Galvani and Volta
remained on relatively good terms throughout the debate. As the
controversy spread beyond Italy into France, Germany, and even
into the halls of the Royal Society in England, Galvani's and
Volta's supporters became increasingly partisan and launched a
seemingly unending series of vitriolic charges and counter charges
at each other. Galvani and Volta, on the other hand, continued to
refer to each other with the greatest respect.

The other interesting thing about the entire dispute is that
they were both right, at least partially. Of course, both men were
also partially wrong. Before their discoveries, electrical enthusiasts had been working with static electricity, similar to what is
produced by scuffing your feet across a rug and then touching
a metal doorknob. What Galvani discovered was current electricity, the stuff we use to run our lights and TVs and computers
and everything else that makes our modern world possible. The
problem was that Galvani didn't recognize what he had discovered and clung to the idea that somehow biological electricity
was different from other electricity. Volta created the battery, but
didn't realize that a chemical reaction was responsible for what
he saw. He thought he had created some sort of perpetual-motion
device.

It would take years for the two of them to get the credit that
they deserved, and in the meantime other events were sweep-
ing across Europe. In the 1790s, while scientific minds were arguing over the intricacies of electricity, Napoleon's army was
marching across the continent. In 1797, only six years after the
publication of his great work, Galvani's beloved Bologna became
part of Napoleon's Cisalpine Republic.

The new authorities demanded that all civil servants take an oath of loyalty to the republic. As a university professor, Galvani was required to take the oath, but he considered it a violation of his political and religious beliefs. He refused. The authorities retaliated by stripping him of his position. Without a job, or any other source of income, he was forced out of his home and had to rely on the charity of his brother.

Desperately, his friends and supporters tried to appeal the decision on his behalf. They argued that he should be granted an exemption because of his fame and scientific accomplishments. Eventually, they succeeded. The exemption was granted, but by then it was too late. Luigi Galvani died almost penniless in 1798, before the exemption could take effect.

Galvani's experiments on the link between electricity and living organisms were significant but might not have become the stuff to inspire mad scientists without the intervention of another scientist, one of his greatest proponents, Giovanni Aldini. He was Galvani's nephew and also a scientist and lecturer of physics at the University of Bologna. He was one of the foremost proponents of galvanism, and formed a society of Galvani supporters to defend his famous relative. What's more, he took his uncle Luigi's ideas on the road.

Giovanni Aldini was born on April 10, 1762, in the city of Bologna, like his uncle. He was the son of Galvani's sister Caterina. At an early age, the boy became fascinated with his uncle's scientific experiments. He assisted in many of those experiments and even helped Galvani edit his 1791 book on electricity. Aldini later went on to formally study physics at the University of Bologna. He graduated, and in 1798, shortly before his uncle's death, he became a physics lecturer at the university.

Aldini continued to explore galvanism. Although, he was a physicist by training, he became interested in the medicinal applications of his uncle's work. He believed strongly that he had firsthand knowledge of its curative powers, and wrote in 1794:

> This sort of revolution in Galvani's system came to my attention at the time he [Galvani, who was primarily a physician and an anatomist] was treating me for a deadly fever. After having escaped, thanks to his generous care and efforts, a nearly unavoidable death, I started to work zealously to bring support to a doctrine that I trusted, despite the attacks under which it came. I felt at ease to be able to pay a tribute to the truth and, at the same time, to provide Galvani with a public account of my gratitude (Parent, 2004).

As part of his search for the medical benefits of galvanism, Aldini expanded on his uncle's research, and began experimenting on warm-blooded animals. He conducted his research on birds, lambs, calves, and oxen. In one series of these experiments, he applied electric current directly to the brain of an ox. He noted that he got a particularly strong response when he applied the electricity to the regions of the brain known as the corpus callosum and the cerebellum. Ironically, to produce the electricity needed for his galvanic experiments, Aldini ended up relying on one of Volta's electric piles.

Later, he wrote that his ultimate goal from this line of research was to use electricity as a means of reanimating and/or controlling the vital forces of living things. This was the great benefit he wished for science to reap from the theory of galvanism.

Once Aldini was satisfied with his results in animals, he began doing experiments on humans. This time, his subjects were the bodies of convicted criminals. He was able to produce all manner of muscular contractions by applying an electric arc to various areas of the head and body. He noted, however, that the bodies had to be fresh, because, the effects faded after a few hours. He was also puzzled and quite disappointed when he noticed that his electrical shocks failed to produce any significant effects in the victim's heart.

Having satisfied himself of the effects of galvanism on the dead, Aldini began experimenting on the living, but first, he tried it on himself. He used a small pile, consisting of only about fifteen of the zinc and copper discs, and described the effects of the first application: "First, the fluid took over a large part of my brain, which felt a strong shock, a sort of jolt against the inner surface of my skull. The effect increased further as I moved the electric arcs from one ear to the other. I felt a strong head stroke and I became insomniac for several days."

Next, Aldini tried to apply galvanism therapeutically. The first patient to receive the new treatment was a twenty-seven-year-old farmer named Luigi Lanzarini. He had been committed to Santo Orsola Hospital in Bologna for what was described as melancholy madness, what today we would call major depression. In the spring of 1801, Aldini began administering a series of shocks to the farmer's head. The results were encouraging. Over the course of the next several weeks, Aldini increased the intensity of the shocks, and the patient experienced a full recovery. Within a few days after the treatments were completed, Lanzarini was sufficiently cured to return to his family.

Now Aldini felt he was ready to reveal the wonders of galvanism to the public. In January and February 1802, he gave his first public demonstrations in an open area near Bologna's Palace of Justice. To the amazement of those present, he applied electric current to the bodies of several freshly executed criminals. As the current ran through various parts of the bodies, it elicited twitches and convulsions from the dead men as well as cheers and applause from the audience.

Aldini's response to the audience's attention was equally strong. He decided to take his show on the road to demonstrate the miracles of electricity to a wider audience. By fall 1802, he was touring Europe. Long before cable TV and video games, people at the turn of the nineteenth century would amuse themselves by attending scientific demonstrations. They paid a small admission fee to witness the latest scientific wonders of the day, and Aldini, now being both a scientist and showman, gave them their money's worth.

He would take the corpses of animals and use the powers of galvanism to reanimate them. For instance, he would often take the severed head of an ox, and with the aid of one of Volta's piles apply electric current to various parts of it. The eyes would flutter open, the face would contort, and the animal's mouth would open and close as if trying to moo. He would conduct similar demonstrations on sheep, horses, and dogs. In the words of one of the actual eye witnesses:

> Aldini, after having cut off the head of a dog, makes the current of a strong battery go through it: the mere contact triggers really terrible convulsions. The jaws open, the teeth chatter, the eyes roll in their sockets; and if reason did not

stop the fired imagination, one would almost believe that the animal is suffering and alive again.

As if that weren't enough to produce a sensation among the audience, he would again perform similar demonstrations upon the bodies of executed criminals. In one famous case in 1802 in England, he used the freshly executed corpse of George Foster, who had been convicted of drowning his wife and children. Aldini used a battery made from several hundred zinc and copper plates, and after moistening the dead man's ears with saltwater, he applied the electrodes. According to news reports in the *Newgate Calendar*:

> On the first application of the process to the face, the jaws of the deceased criminal began to quiver, and the adjoining muscles were horribly contorted, and one eye was actually opened. In the subsequent part of the process the right hand was raised and clenched, and the legs and thighs were set in motion. Mr. Pass, the beadle of the Surgeons' Company, who was officially present during this experiment, was so alarmed that he died of fright soon after his return home.

These demonstrations electrified not only the bodies of the dead, but the imaginations of the living. Word, as well as controversy, swept through Europe. News reports appeared across the Continent, as other researchers tried to duplicate the results, and journals of the day argued the scientific merits and ethical implications of such work. Needless to say, this was a hot topic of discussion in the bleak summer of 1816, when the newlywed Mary Wollstonecraft Shelley, her husband, Percy Bysshe Shelley, and their companions visited the villa of Lord Byron. To amuse themselves while cooped up inside by the cold and stormy weather, they competed to create spooky stories. The young Mary trumped the more experienced writers in the group. The result was the novel *Frankenstein; or, The Modern Prometheus*. In her introduction to the 1831 edition of the book, Mary Shelley recounts some of the long scientific conversations between her husband and Lord Byron that inspired her:

> They talked of the experiments of Dr Darwin. . . . who preserved a piece of vermicelli in a glass case till by some extraordinary means it began to move with a voluntary motion. Not thus, after all, would life be given. Perhaps a corpse would be reanimated; galvanism had given token of such things; perhaps the component parts of a creature might be manufactured, brought together, and endued with vital warmth.

The reference to Darwin refers not to Charles Darwin. His revolutionary theories would not be published until 1859. It is a reference to his grandfather Erasmus Darwin, a well-known naturalist in his own right, member of the Lunar Society, and friend of both Joseph Priestley and Benjamin Franklin. He conducted experiments to replicate the work of Galvani and Aldini. Thus, the work of these real scientists and their daring efforts to explore the phenomenon of electricity may have directly contributed to *Frankenstein,* the seminal work of science fiction that inspired the image of the mad scientist for generations to come.

The next, some say greatest, contributor to the legend of the mad scientist would be born half a century after Aldini animated the corpses of the dead. In his time, he would render spectacles for the public even more electrifying. He would lay the foundation for our modern world, and generate even more fame and controversy for himself and his inventions than most men can even dream of. To reach his goals, he would navigate the high finance world of the robber barons, hobnob with literary and artistic notables such as Mark Twain and Sarah Bernhardt, and do battle with his arch nemesis, the Wizard of Menlo Park. To this day, his name is capable of eliciting awe, astonishment, and arguments. That name is Nikola Tesla.

In the midst of a violent thunderstorm, at precisely the stroke of midnight on June 28, 1856 (July 10 by the Gregorian calendar) Nikola Tesla came into the world. He was born in the Croatian village of Smiljan, then part of the Austro-Hungarian Empire, the fourth child of his Serbian parents. Unlike Galvani, Tesla's father, Milutin Tesla, was a clergyman, and encouraged his son to go into the Serbian Orthodox Church. At that time, the other acceptable option for a young man of his social position was a career in the military. Tesla was not inclined toward either path. Instead, he showed from an early age a fascination with inventing.

Tesla's mother, Đuka Mandic Tesla, was a beautiful and gifted woman, who if she had been born in a different place and a different time, might have been a successful inventor herself. In later life, Tesla always credited her with being the source of his genius. As an adult, he wrote of her, "She invented and constructed all kinds of tools and devices and wove the finest designs from thread which was spun by her. She even planted the seeds, raised the plants, and separated the fibers herself. She worked indefatigably, from break of day till late at night, and most of the wearing apparel and furnishings of the home was the product of her hands" (Tesla, 1919).

Tesla's mother always encouraged her precocious young son. When he was only five years old, he built a small paddleless waterwheel. It was quite unlike any of the waterwheels he would have seen growing up, but when he placed it in the moving waters of a small stream near his home, it turned smoothly. This would mark the beginning of Tesla's fascination with harnessing the power of moving water. After reading descriptions of Niagara Falls a few years later in school, he imagined a big wheel turning in its cascading waters. He told his uncle that one day he would go to America and build it.

He was successful in school and when not studying would continue his youthful engineering. Shortly after making plans for the mighty Niagara Falls, he built a motor out of splinters and small pieces of wood that was powered by harnessed June bugs. Of course, not all of his engineering endeavors where quite so successful. He learned about gears and springs by taking apart and then reassembling several of his grandfather's clocks. In Tesla's own words, "In the former operation I was always successful but often failed in the latter."

Despite his youthful accomplishments and long-lasting desire to study engineering, Tesla's father was dead set on his son becoming a priest. That might have been his fate, if the boy had not contracted cholera. The disease ravaged the teenager and for months, he lay in bed, too weak to rise. For much of that time, his father sat by the bedside praying for the recovery of his son. At one point in the illness, Tesla turned to his father, and weakly said, "Perhaps I may get well if you will let me study engineering." Desperate for a way to cheer his son, the father agreed. Months later, when Tesla recovered, his father kept the promise.

Tesla traveled to the Austrian city of Graz and enrolled in the Austrian Polytechnic to study electrical engineering. It was there that he became intrigued with alternating current. Up until then, electrical motors and other electrical devices ran on direct current (DC), where the electrical current flows directly from the generator to the machine doing the work, called the *load*.

One day, one of Tesla's teachers, Professor Poeschl demonstrated to his class something called a Gramme machine, a device that can be operated as a motor or generator. Like all such devices of the time, it ran on direct current. As the professor worked the machine, it sparked badly. Tesla, who had been examining the machine closely, suggested that it might be improved by converting it to alternating current (AC). With alternating current, electricity flows from the generator to the load and then back again in a series of alternating cycles. Poescl derided the young student and responded, "Mr. Tesla may accomplish great things, but he will never do this." He continued, "It is a perpetual motion machine, an impossible idea." Despite his professor's predictions, Tesla did go on to design, build, and patent the world's first practical AC motor. The way that he did so may also have been the start of his reputation as a mad scientist.

Tesla continued to be plagued by health problems even after he graduated. In 1881, while working for the Hungarian Central Telegraph Office in Budapest, he suffered an unusual affliction that his doctors were unable to diagnose. His heartbeat fluctuated wildly between abnormally low and alarmingly high. He reported that all of his senses seemed to become hyperacute. He could hear a watch ticking three rooms away. A passing carriage caused his whole body to shake, and the pain caused by a train whistle twenty miles away was almost unbearable. He described exposure to the sun's rays as delivering stunning blows, and minor vibrations in the ground caused him such distress that he was unable to rest without rubber cushions being installed under his bed. These symptoms were so severe that he did not expect to recover.

However, as quickly and mysteriously as these afflictions appeared, they seemed to disappear. That is, until Tesla and one of his close friends were going for a long walk in one of the city's parks near dusk. As the sun sank toward the horizon, Tesla and his friend were discussing Goethe's great work *Faust,* when Tesla seemed to be seized with some sort of fit. He froze with his long arms extended. His alarmed friend tried to lead him to a nearby bench, but Tesla resisted. He found a stick, and began frantically drawing in the dirt. He later reported, "The idea came like a flash of lightning, and in an instant the truth was revealed." Tesla's wild scratchings in the dirt were actually a diagram for a completely new type of AC motor.

Other inventors and engineers had designed AC motors before, but none was terribly successful. Previous attempts had tried to design an AC motor using a single circuit, similar to a DC motor. Tesla's system used multiple circuits that were powered in alternating series to create a rotating magnetic field. This was a vast improvement and would lay the groundwork for today's modern electric distribution system.

Unfortunately, Tesla had no money to build and test his revolutionary new motor. Instead, in 1882, he traveled to Paris to take a job with Thomas Alva Edison's French telephone subsidiary. He worked hard and used his engineering talent to impress his employers. Soon they were giving him assignments to act as a troubleshooter, solving problems in Alsace and then Strasbourg. One after the other, he found ingenious solutions to the problems and even managed to build a prototype of his AC motor in his spare time.

One of Tesla's managers was Charles Batchelor, who had been a friend and assistant of Edison for many years. Batchelor was so impressed by the young Serb that he suggested he might go to America. There, he assured Tesla, the pastures were greener and his prospects would be greater. Tesla agreed. He took what little money he had and bought passage on the ship *Saturnia*. In 1884, with empty pockets and a head full of brilliant dreams, Nikola Tesla sailed to America.

Tesla arrived on U.S. shores, at the age of twenty-eight, with an idea, a new bowler hat, and four cents in his pocket. He also carried with him a letter of introduction to the Wizard of Menlo Park, the greatest inventor in America, Mr. Edison himself. Batchelor had written his young friend a glowing letter of recommendation to his former boss. It contained the following, "I know two great men and you are one of them; the other is this young man!"

Edison initially seemed unimpressed, and he wanted nothing to do with any AC schemes, which he considered far too dangerous. However, he was in desperate need of a new engineer, so he gave Tesla a job. The young Serb took full advantage and used the new position to demonstrate his talent, dedication, and predisposition toward hard work. He had difficulty admitting it, but Edison was impressed.

The two men were both hardworking geniuses, but other

than that, they could not have been more different. They seemed
to attack problems from very different angles. Tesla was as much
a theoretician as he was an engineer. Edison was strictly practical with no time to waste on theory. Edison was rather careless
about his personal appearance and hygiene. Tesla was an impec-
cable dresser with a phobia of germs. Tesla was educated and
rather sophisticated. Edison was largely self-taught, and once
when unable to locate the Serbian's obscure birthplace on a map,
actually asked Tesla if he had ever tasted human flesh.

In spite of their differences, Edison recognized that he could
use Tesla. The American was then supplying DC power for much
of New York City. However, he had been having a series of problems keeping the dynamos he used to generate the power running
properly. Reluctantly, he gave Tesla the assignment of improving
the dynamos. If he succeeded, Edison offered him the princely
sum of fifty thousand dollars.

Tesla threw himself wholeheartedly into the challenge. He
worked long hours with very little sleep for months on end to
improve Edison's twenty-four dynamos. He eventually succeeded
not only greatly improving their efficiency but also in installing
a new automatic control system, for which he later received patents. He returned to Edison, showing off his work and expecting
the payment that he had been promised. Edison laughed at him
and said, "Tesla, you don't understand our American humor."
Instead of the fifty thousand dollars, Edison offered him a ten
dollar raise. Tesla promptly quit, never to return.

Fortunately for Tesla, a group of investors got wind of the
young inventor's reputation. They offered to finance a company
in his name. The Tesla Light Company was born. It was headquartered in Rahway, New Jersey, and its first big job was to

improve the arc lights that were currently in use. The arc light, which had originally been developed by Humphry Davy, was starting to be used to light streets. Tesla quickly devised an improved version, and it was soon adopted to light the streets of Rahway. It was quite successful, but unfortunately, unlike the long-term vision of Tesla, the investors were mainly interested in short-term profit. Once they had used him to make their money, they eased Tesla out of the company. All he ended up with were some nicely engraved but worthless stock certificates.

At that time, in 1886, America was in the midst of a recession, and jobs where scarce. Tesla held seven patents at the time for his improved control system and arc lights, but the ambitious young inventor was forced to subsist on manual labor jobs. For months he toiled as a laborer, but eventually, the foreman on his crew took pity on him and introduced Tesla to the manager of the Western Union Telegraph Company, A. K. Brown. As it turned out, Brown knew about alternating current, and he too recognized Tesla's talents. For the second time in as many years, the young immigrant received financing to start his own company, this time the Tesla Electric Company.

In April 1887, the new company opened for business at 33-35 South Fifth Street, just blocks from Edison's workshops. Tesla was soon producing AC systems—single-phase, two-phase, and three-phase, and for each type he developed and built the dynamos, transformers, motors, and automatic control systems to go with them. Between 1887 and 1891, Nikola Tesla received forty patents and was soon providing some serious competition to Edison's DC systems.

AC power has a number of advantages over DC. For one thing, direct current cannot be efficiently transmitted over long distances. Edison's solution was to install power stations every mile or so. The price of real estate in New York made this expensive, and it was totally impractical in rural areas. Another disadvantage is that DC generators cannot change their voltage, the unit of electricity named after Volta. This means that you need one set of generators and transmission wires for homes and another for factories that need higher voltages. In New York, this required a system of heavy wires that made parts of the city look like they were trapped in a giant spiderweb.

AC power, on the other hand, is much easier to transmit over long distances. It is also possible to use a device called a transformer to change the voltage, so today homes can have 120 volts and factories can have 240 volts, using the same generators and transmission lines. The lines themselves can also be made thinner, using less copper and making them much cheaper to install.

In 1888, Tesla was invited to deliver a lecture on the benefits of alternating current before the American Institute of Electrical Engineers. One of those present at the lecture, Dr. B. A. Behrend, later commented, "Not since the appearance of Faraday's 'Experimental Researches in Electricity' has a great experimental truth been voiced so simply and so clearly. . . . He left nothing to be done by those who follow him. His paper contained the skeleton even of the mathematical theory."

Word of Tesla's success and his new and improved system soon spread through Wall Street. There it came to the attention of George Westinghouse. Westinghouse was an inventor himself, who had developed an improved type of air brake for trains. He had parlayed his inventions and investments into an empire of business interests and, in the process, become one of the most influential businessmen in America. He had long been interested in AC power, even investing in one of the earlier systems before Tesla's improved version came along.

The walrus-mustached business magnate went to Tesla's lab to see for himself what the young upstart had come up with. Within moments of his arrival, Westinghouse was craning and bending to get a better look at the wondrous machines and gadgets that filled Tesla's lab. The businessman was soon asking the young inventor questions about this device or that. Tesla was impressed with the intelligence of the questions. The two men had a natural affinity for one another and quickly became friends.

Almost immediately, Westinghouse saw these advantages of Tesla's AC system. He offered Tesla $60,000 for his patents, including $5,000 in cash and 150 shares of Westinghouse stock. As amazing as that amount of money was back then, the offer also included a royalty of $2.50 per horsepower for the electricity sold. Within a very few years, a royalty like that would be worth a mind-boggling amount of money. Potentially, it could make Tesla one of the richest inventors in the world.

Once the documents were drawn up and the ink on the signatures was dry, Westinghouse set about promoting Tesla's new system for AC power. He hired Tesla to work as a consultant on the installation of the system, paying him a salary of two thousand dollars per month. Of course, since Westinghouse's operations were headquartered in Pittsburgh, it required that Tesla move to the city, but that was a small price to pay to see his dream of AC power become a reality.

In the midst of his work for Westinghouse, Tesla achieved what he described as his greatest accomplishment. On July 30, 1891, Nikola Tesla became a U.S. citizen. In later years, he often told friends that he valued that more than all the scientific honors that were bestowed upon him. Evidence for that is provided by the many honorary degrees which he tossed carelessly into drawers, while his certificate of naturalization, Tesla always kept secure in his office safe.

Thomas Edison was outraged when he heard of Tesla's deal with Westinghouse. Here was a real threat to his DC monopoly and a threat that the Wizard of Menlo Park had no intention of taking lying down. The battle lines were drawn. While Tesla had Westinghouse on his side, Edison had the backing of J. P. Morgan, one of the world's most powerful bankers and a man whose influence reached all corners of commerce and government. What ensued would come to be known as "The War of the Currents." The stakes were the chance to harness the power of Niagara Falls.

An international committee was convened to award the contract for a system that would use Niagara Falls to generate electricity. Both sides relished that prize. While Westinghouse began to publicize the advantages of AC, Edison responded by launching a publicity campaign of his own about its dangers. He was convinced because of the higher voltages possible with AC that it was inherently unsafe. To convince others, he hired local boys to capture neighborhood cats and dogs, which he then proceeded to electrocute with AC before on-looking reporters. Edison would then urge them to report on the dangers of people being electrocuted, or "Westinghoused," as he called it.

These brutal demonstrations escalated over the course of years to include calves, horses, cattle, and ultimately even an unfortunate elephant. In 1903, officials at the Luna Park Zoo in Coney Island decided that Topsy the elephant had become so dangerous that she needed to be euthanized. Over the course of three years, she had killed three of her handlers. Never mind that at least one of them was later discovered to have tried feeding her a lit cigarette. Once the decision was made, a means of dispatching the beast was sought. Edison leaped at the chance.

On January 4, 1903, before an audience of fifteen hundred spectators and reporters, the condemned pachyderm was prepared. Copper electrodes were attached to her feet and wires were run to a generator. When the technicians gave the signal, the switch was thrown and the sixty-six-hundred-volt AC charge ran through Topsy's body. As onlookers gasped, smoke rose from her feet, and the mighty elephant keeled over dead. Edison's assistants filmed the entire spectacle, and the film was later released to the public.

If that weren't enough to terrify the public, one of Edison's assistants, Harold P. Brown, had previously developed the ultimate device to demonstrate the horrors of alternating current, the electric chair. The New York State Legislature had shown interest in using electricity to execute convicted criminals, but neither Edison nor Westinghouse wanted their system associated with human executions. However, Edison and Brown managed to covertly obtain several of Westinghouse's AC generators, and used them for their execution device.

William Kemmler, a convicted murderer, was scheduled for execution at Sing Sing Prison on August 6, 1890. While the seventeen witnesses watched, electrodes were attached to Kemmler's body, and he was strapped into the chair. His last words as they put the hood over his head were, "Take it easy and do it properly, I'm in no hurry." When the order was given, the prisoner received an AC shock of a thousand volts. He was pronounced dead by the physician present, Dr. Edward Charles Spitzka, but several witnesses said that they saw him breathing.

Once Dr. Spitzka confirmed that Kemmler was still alive, he cried out, "Have the current turned on again, quick—no delay!" This time they gave the prisoner two thousand volts. Several of the blood vessels under his skin ruptured, and before the horrified witnesses, Kemmler began to bleed. Several of the witnesses claimed that he caught fire, and reported the smell of burned flesh. It took eight minutes to complete the execution, and one of the reporters present described it as, "an awful spectacle, far worse than hanging."

That horrifying scene might have ended the AC versus DC debate, but Westinghouse had a few tricks up his own sleeve. He conceived of a way to demonstrate the wonderful utility and safety of alternating current before the entire world. His scheme unfolded as plans were made for the 1893 Chicago World's Fair. The Columbia Exposition, as it was called to commemorate the four hundredth anniversary of Columbus' discovery of the New World, was to be the first electric world's fair. Edison made a bid to provide all of the electricity and lights for one million dollars. Westinghouse underbid him by half. With that masterful stroke, he secured for himself and Tesla just the public platform they needed. It would become the decisive battle in The War of the Currents and enshrine Tesla forever among the pantheon of mad scientists.

On May 1, 1893, President Grover Cleveland turned the gold key that raised flags, switched on fountains and lit one hundred thousand light bulbs to herald the opening of the exposition, all of it powered by alternating current. In the course of six months, the fair would welcome twenty-five million visitors, then approximately one third of the U.S. population. They were treated to the world premiers of the Ferris wheel, Cracker Jack, and the exotic dancer Little Egypt, but electricity was the undisputed star of the show. In the Hall of Electricity, Tesla, adorned in white tie and tails and protected by thick cork-soled shoes, showcased his electrical wonders before one of his newly developed neon signs proclaiming, "Welcome Electricians."

He presented to the public an egg-shaped copper ball, dubbed the Egg of Columbus that he induced to spin, as if with a life of its own, using rotating magnetic fields. They saw induction motors, transformers, switchboards, and polyphase generators. The dashing inventor took glass tubes, the forerunners of today's fluorescent lights, and caused them to glow without wires, simply by wielding them like swords within an electrical field. All of this was accompanied by two insulated plates from which issued lightning-like electrical discharges and claps of thunder. His demonstrations were awe inspiring as well as death defying. As one of the many newspaper journalists present reported:

Mr. Tesla has been seen receiving through his hands currents at a potential of more than 200,000 volts, vibrating a million times per second, and manifesting themselves in dazzling streams of light. . . . After such a striking test, which by the way, no one has displayed a hurried inclination to repeat, Mr. Tesla's body and clothing have continued for some time to emit fine glimmers or halos of splintered light. In fact, an actual flame is produced by the agitation of electrostatically charged molecules, and the curious spectacle can be seen of puissant, white, ethereal flames, that do not consume anything, bursting from the ends of an induction coil as though it were the bush on holy ground.

By the time the fair was over, the battle to see who would win the contract to harness Niagara Falls for hydroelectric power had been won, and Tesla's fame was ensured. It was a monumental project, taking years to complete, but on November 16, 1896, the switch was thrown and Tesla's AC power lit up the nearby city of Buffalo, New York. He quickly became the toast of high society and one of the most famous men in the world. He undoubtedly would have been one of the richest as well, but alas, Tesla's business acumen was not the equal of his inventive genius or his loyalty to good friends.

Even though he was victorious, The War of the Currents cost George Westinghouse dearly. J. P. Morgan had pulled out every nasty trick in the book, spreading rumors on Wall Street of Westinghouse's mismanagement and outright incompetence. This severely devalued Westinghouse stock. Morgan then exerted his considerable power with his fellow bankers to deny Westinghouse loans and put pressure on his company. To survive, Westinghouse was forced to arrange a merger with several smaller companies, including U.S. Electric Company and the Consolidated Electric Light Company. The goal was to form a new corporation to be called the Westinghouse Electric and Manufacturing Company that would be larger and less vulnerable to Morgan's manipulations.

Unfortunately, Westinghouse's partners in the merger balked when they learned of the value of Tesla's royalties. Within a few years of signing the contract, the $2.50 per horsepower royalty accorded the inventor was estimated to have accrued a value in excess of $12 million. Add to that the money from motors, powerhouse equipment, and all the other applications of AC power, and Tesla stood to become one of the world's first billionaires. It was enough to sink the hope of any merger deal. If he were to save the company, Westinghouse would have to get rid of his royalty contract to Tesla.

Being an inventor at heart himself, Westinghouse was loath to do so, but under pressure from his bankers and investors, he had no choice. Reluctantly, he approached his Serbian friend and explained the situation. Tesla asked what would happen if he refused, and Westinghouse bluntly told him it would be the end of the Westinghouse Company. It would also mean the end of his dream of offering AC power to the people of the world.

Tesla was always first and foremost concerned with his inventions. He enjoyed the high life and fine clothes and dining, but to him, money was simply a means to an end, not the goal itself. At last, he turned to his industrialist friend and said:

Mr. Westinghouse, you have been my friend, you believed in me when others had no faith; you were brave enough to go ahead . . . when others lacked courage; you supported me when even your own engineers lacked vision to see the big things ahead that you and I saw; you have stood by me as a friend. . . . You will save your company so that you can develop my inventions. Here is your contract and here is my contract—I will tear both of them to pieces, and you will no longer have any troubles from my royalties. Is that sufficient? (O'Neill, 1944).

With that, the inventor tour up the contracts worth a fortune in exchange for $216,600 for the outright purchase of his patents. Without the threat of prohibitive royalties, the mergers went through in 1889. The company was saved.

It would have been easy for him to sit on his laurels, but that is not the stuff that true geniuses are made of. Tesla unveiled his next wonder at Madison Square Garden in 1898. He demonstrated for the public, in an indoor pond created for the event, the world's first remotely controlled vehicle. It was a boat controlled by short range radio signals. Tesla described the reaction, "When first shown . . . it created a sensation such as no other invention of mine has ever produced." The crowd jockeyed to see as Tesla sent the little iron-hulled craft this way and that.

Some accused him of having a midget secreted on board to control it. When Tesla showed them they were wrong, others began to suspect that the inventor was controlling it all with his mind. A *New York Times* report suggested that it would make an effective weapon, if Tesla could make it submerge and then filled it with explosives. The inventor angrily replied, "You do not see there a wireless torpedo, you see there the first of a race of robots, mechanical men which will do the laborious work of the human race." Unfortunately few at the time could appreciate the practical application of such a device, and Tesla was unable to market it.

Around this time, one of Tesla's more amusing experiments took place. He had long been interested in the phenomenon of resonance. This can be demonstrated by pushing someone on a swing. By applying a very small push at just the right time as the person swings back and forth, it is possible to send them higher and higher in the air. That's what Tesla was investigating one day in his lab at 46 East Houston Street. The building had a large iron pillar that passed through the lab and down into the building's foundation beneath. Tesla attached a small mechanical oscillator to the beam, and began adjusting the frequency.

To his delight, Tesla found that by tuning it to just the right frequency he could make various pieces of equipment and furniture in the lab vibrate and shake across the floor. Unbeknownst to Tesla, the vibrations from the oscillator were traveling down the pillar and into the surrounding understructure below the building. While the inventor calmly continued his experiment, reports started coming into the local police station in Manhattan that windows were shattering, and people were fleeing their homes and businesses in the belief that an earthquake had struck New York.

When the police managed to localize the area of the disturbance, and noticed that the lab of a certain notorious genius was at the exact center, they dispatched a pair of officers to investigate. Meanwhile, Tesla had begun to sense that the vibrations from his device were becoming ominously strong. He decided to end the experiment, and in his haste, smashed the oscillator with a hammer. Just then the two officers burst through his door. Tesla calmly turned to them and said, "Gentlemen, I am sorry. You are just a trifle too late to witness my experiment. I found it necessary to stop it suddenly and unexpectedly and in an unusual way" (O'Neill, 1944).

Once the furor had died down, Tesla redirected his interest in frequencies. Like many scientists at the time, Tesla became fascinated with using higher and higher frequencies of electricity. It offered a number of technical advantages but was difficult to achieve. Tesla tried using conventional AC generators, running them at higher and higher frequencies, but as they approached twenty thousand cycles per second, the machines would violently shake themselves apart.

To solve the problem, the inventor created a device still known as a Tesla coil. The ingenious contraption was capable of taking ordinary household current, running at sixty cycles per second, and stepping it up to extremely high frequencies, as much as hundreds of thousands of cycles per second. Among the dazzling array of technical achievements he could accomplish with his coil, such as broadcasting radio waves over long distances, Tesla found that he was able to transmit electricity to a fluorescent tube, and make it glow brightly without benefit of wires. From that point on Tesla became obsessed with the possibility of transmitting electricity through the air.

While Tesla began experimenting with transmitting electrical power, an Italian inventor named Guglielmo Marconi, was trying to broadcast radio signals. He managed to achieve wireless communication and filed for a patent in England in 1896, but Marconi's system was capable of sending signals only over very short distances. Later, he managed to demonstrate that he could broadcast signals across the English Channel, but ironically, he needed a Tesla oscillator to do it.

Tesla himself had been sending signals between his lab in New York and West Point, fifty miles away, as early as 1895, but around that time a disastrous fire destroyed his lab and much of his work. A great deal was lost forever, but Tesla was able to reconstruct his radio equipment and filed for American patents for basic radio in 1897. When Marconi tried to file for American patents in 1900, the Patent Office turned him down, citing Tesla's precedent. It would be the beginning of another bitter battle.

Marconi had powerful connections in England, and equally powerful financial backers in America. His investors and supporters included not only Edison but Andrew Carnegie, the steel magnate. When asked if he was worried about the Italian inventor winning, Tesla calmly responded, "Marconi is a good fellow. Let him continue. He is using seventeen of my patents."

Unfortunately, Tesla's confidence was premature. Marconi with help from his supporters behind the scenes, managed to get the U.S. Patent Office to reverse itself. In 1904, they awarded Marconi the patent for radio. The final insult came when Marconi was awarded the Nobel Prize in 1911 for the invention of radio. Tesla was outraged. He sued the Marconi Company for infringement of his patents in 1915. The case ended up being dragged out for decades and would end in the Supreme Court. They would eventually rule in Tesla's favor but not until after the inventor's death in 1943.

While all of this was going on, Tesla continued to pursue his dream of broadcasting electrical power. To free himself from the distractions of New York, he moved his research to a secluded location in Colorado Springs, Colorado. Against the mountain backdrop, he constructed a laboratory worthy of any mad scientist, complete with fences and signs warning would-be visitors: KEEP OUT—GREAT DANGER. If that were not sufficient, the front gate had posted a quote from Dante's *Inferno,* "Abandon hope all ye who enter here."

When it was completed and fully operational, the lab contained banks of instruments and control panels as well as a giant version of his Tesla coil. Above the lab, a mast towered over 200 feet and lit up the night sky with electrical discharges to rival the wildest mountain thunderstorms. In one of his more spectacular tests on a night in 1899, Tesla generated lightning bolts that shot 135 feet from the mast's copper tip. Witnesses in Cripple Creek, fifteen miles away, reported hearing the crack of the thunder being produced. Unfortunately, at the height of the experiment everything suddenly stopped. Tesla's electrical spectacle had accidentally overloaded and set fire to the dynamo that supplied power to the nearby town. All of Colorado Springs was plunged into darkness.

Once Tesla had repaired the dynamos, he was back at work. Late one night, while conducting research in his laboratory, Tesla began detecting strange signals. They were oddly repeating radio pulses. Not knowing what to make of it, Tesla thought that they were radio signals being sent from outer space by Martians. When he reported his discovery to the media, it caused a sensation. Some believed that we had achieved communication with an otherworldly civilization, others accused Tesla of staging a hoax, still others thought that the world famous genius had gone mad. The radio signals from outer space were genuine, but what no one, including Tesla, realized at the time was that they were given off by the stars themselves. Nikola Tesla had inadvertently become the world's first radio astronomer.

From this point on, his reputation as a mad scientist only grew. By the time the twentieth century began, Tesla had already laid the basic groundwork for radio transmission, x-rays, electron microscopes, radio astronomy, neon and fluorescent lights, and remotely controlled vehicles. He would go on to develop a bladeless turbine and file patents that contained the basic principles underlying computer circuits. In 1917, in what many consider to be the ultimate in irony, he was awarded the Thomas Edison Medal by the American Institute of Electrical Engineers.

However, even as he continued to come up with ideas, Tesla fell on hard times. His behavior had always been eccentric. He had a number of phobias, including fears of germs, jewelry, and contact with other people's hair. He also suffered from what would now be called obsessive-compulsive disorder. While eating, he could not enjoy the meal until he had calculated the cubic volume of all of the food. He was obsessed with the number three. For instance, if he walked around the block, he would feel compelled to do so three times. When he was rich and famous, these things were dismissed as amusing quirks, but as his money began to fail after a series of bad investments and unsuccessful inventions, they began adding to his reputation as a mad scientist.

He did little to downplay this reputation. He simply went on with his work and continued to make grandiose claims about the future. In a *New York Times* article published in 1933 he went so far as to propose building the ultimate toy of mad science, the death ray. He described for the interviewer a series of towers that "will send concentrated beams of particles through the free air, of such tremendous energy that they will bring down a fleet of 10,000 enemy airplanes at a distance of 250 miles from the defending nation's border and will cause armies of millions to drop dead in their tracks."

Although he never actually built such a device, rumors continued to spread that he had. Within days after his death in 1943, the FBI searched and confiscated all of his papers. The seized materials were immediately classified. Shortly after that, Tesla's laboratory mysteriously burned to the ground.

By any accounting, Nikola Tesla was a remarkable figure. During his lifetime, he created means of making our modern world possible. AC power, fluorescent lights, robotics, computers, radio telescopes, electron microscopes, particle beam weapons, and much, much more can trace their origins to this singular mind. The best epitaph for such a visionary may best be found in his own words, "Let the future tell the truth, and evaluate each one according to his work and accomplishments. The present is theirs; the future, for which I have really worked, is mine."

CHAPTER SIX

Atomic Secrets

Unleashing the Power of Physics

FOR CENTURIES, THE ALCHEMISTS TOILED TO UNLOCK the secrets of matter for the promise of near limitless power. As the nineteenth century waned and the twentieth dawned, the physicist would take up the challenge. In the early 1800s, John Dalton managed to pull together various discoveries about matter into a cohesive theory of the atom. Others used his work to discover a new array of chemical elements, and by the second half of the century, Dimitri Mendeleev would organize them into the periodic table. This laid the groundwork for the discoveries to come.

The next set of breakthrough discoveries came at the hands of someone who at first seems to defy the image of mad scientist but would ultimately come to exemplify it. Although we often think of the mad genius as male, a woman would make significant contributions to the stereotype. Although we often think of the mad genius as solitary, she would go on to form one of the most dynamic and passionate partnerships in the history of science, and although she was born Marya Salomee Sklodowska, she would forever go down in history as Madam Curie.

For a period, the country of Poland disappeared, erased from the maps by its Russian, Prussian, and Austrian occupiers, but on November 7, 1867, a girl named Marya Sklodowska was born in the Russian-controlled Polish city of Warsaw. She was the youngest of five children, but she would someday put her homeland back on the map in her own unique way. Both her parents were teachers. Her mother, Bronislava, was headmistress of the Freta Street School, the only private school for girls in Warsaw. Marya's father, Wladyslaw Sklodowski had been a physics professor, but since the Russians stripped the Poles of the privilege of teaching physics or chemistry in universities, he was forced to teach at one of the Russian-controlled gymnasiums.

Marya, or Manya as her family called her, was an intelligent child and showed an interest in science. As a toddler, she would often spend time staring at her father's scientific instruments, such as his scales, glass tubes, and prized gold-leafed electroscope locked safely inside their glass cabinet. Both of her parents encouraged their daughter's interest, but when Manya was four, her mother became seriously ill.

Bronislava suffered from consumption, today known as tuberculosis. As the hacking cough and persistent weight loss weakened her, there was little that the doctors could do, other than prescribe rest and extended retreats to milder climates. The family scraped together enough money to send her to a spa in the

Austrian Alps in a desperate hope that the curative powers of the mountain air could help. For two years, Marya and her brother and sisters were denied the company of their mother, and even when Bronislava returned, she would be ever mindful of spreading the disease to her family, eating on separate dishes. Even the comfort of her mother's kiss was denied to the young Marya. By the time she was ten, her mother was dead.

In spite of, or perhaps because of her tragic home life, Marya always did well in school. She and her older sisters attended the Freta Street School where their mother had taught. All of them did well, but Marya proved to be the academic star of the family. Her father decided to provide his daughter with a more challenging environment and enrolled her in the Russian gymnasium. Although she chafed under the strict Russian supervision, Marya again did exceptionally well and graduated at the top of her class.

Now she was faced with a dilemma. While her friends were busy seeking husbands and settling down, Marya and her older sister Bronya had aspirations of higher education. The Russian authorities, however, allowed no further education for Polish women. In response, a number of Polish nationalists organized an underground academy for women. It met in secret, and would come to be known as the Flying University. Marya and Bronya were among the thousand women enrolled in its classes. The two sisters once again excelled and were soon tutoring the other girls.

By now the girls had big plans. Marya wished to be a scientist and Bronya intended to become a doctor, but both of these dreams were impossible in Poland. Their only hope lay beyond their homeland's borders. The Sorbonne in Paris admitted women, but how could they possibly get the money to go there?

Once more, Marya's cleverness served her well. She proposed a bold plan. She would work as a governess and forward the money she earned to Bronya, who could then afford to go to the Sorbonne. When she graduated, and became a doctor, she could then send for Marya and pay for her to attend school.

It took years, but eventually the plan worked. Bronya completed medical school and became a doctor. In 1891, she sent for her little sister. Marya packed up her few possessions and boarded the train for the thousand-mile journey to Paris. She was now twenty-three, and after a lifetime of living under iron-fisted Russian rule, the French metropolis seemed like a marvelous dream. She threw herself into her new life, and when she signed the registration papers at the Sorbonne, she used the French form of her name, Marie.

She lived with her sister for a few months but eventually found that it distracted her from her studies, so Marie took what little money she had and rented a small garret room in the Latin Quarter. Years later, she would write of this period in her autobiography and describe the climb up six flights of stairs to a room so cold that she would often sleep with all of her cloths piled on the bed for warmth. Her meager diet consisted of tea, chocolate, bread, and fruit, with a very occasional bit of meat or perhaps an egg. In spite of this, she described it as, "One of the best memories of my life." She reveled in the academic freedom and the privilege of being taught by some of the best instructors in the world. In her own words:

> All my mind was centered on my studies. I divided my time between courses, experimental work, and study in the library.

In the evening I worked in my room, sometimes very late into
the night. All that I saw and learned was a new delight to me.
It was like a new world open to me, the world of science
which I was at last permitted to know in all liberty (Curie).

She thrived in this environment, and despite being one of only
two women at the Sorbonne seeking degrees in science, Marie
scored at the top of her class. Her success was so inspiring for
other female students in Poland that she was awarded the Alex-
androvich Scholarship for talented students studying abroad.
The six hundred rubles it provided allowed Marie to continue
her studies. Her only disappointment was that in 1894 when she
received her degree in mathematics she ranked only second out
of the entire class. She resolved to do better.

One of her professors, Gabriel Lippmann, a future Nobel
Prize winner, was so impressed that he recommended Marie for
a job investigating the magnetic properties of various types of
steel. The task, however, proved daunting because of the clumsi-
ness of the instruments she was forced to use. When she expressed her frustration, one of Bronya's friends suggested she
consult one of the physicists at the nearby School of Industrial
Physics and Chemistry (EPCI). She recommended a young physicist with a reputation for making very precise instruments
named Pierre Curie.

Thirty years later, Marie recounted their first meeting:

Upon entering the room I perceived, standing framed by the
French window opening on the balcony, a tall young man
with auburn hair and large limpid eyes. I noticed the grave

and gentle expression of his face, as well as a certain detachment in his attitude, suggesting the dreamer absorbed in his reflections. He showed me a simple cordiality and seemed to me very sympathetic. After that first meeting, he expressed the desire to see me again and to continue our conversation of that evening on scientific and social subjects, in which he and I were both interested, on which we seemed to have similar opinions.

Thus began one of the most extraordinary relationships in the history of science. Pierre had already begun making a name for himself in physics, developing a principle of magnetic bodies and temperature that came to be known as Curie's law. He and his brother Jacques, also discovered that certain crystals create electricity when under stress. They named the phenomenon piezoelectricity. Pierre had also designed and built a new type of device called a quadrant electroscope. It could precisely measure minute changes in electric charge and was exactly what Marie needed for her work. Over the course of the next several months, Pierre taught Marie how to properly use the new tool. The two spent many hours together talking of their mutual love of science. On July 26, 1895, they were married.

After they returned from their bicycle tour honeymoon of Brittany, Pierre secured some laboratory space for Marie at the EPCI. Now Marie Curie, she continued her work on magnetic properties, for which she would later receive the Gegner Prize from the Academy of Sciences. She also took classes at night and learned to maintain a household. In the midst of all this, she became pregnant, and on September 12, 1897, gave birth to a daughter. She and Pierre named the girl Irene.

Under the cumulative effects of work and family, Marie began to suffer from serious depression. This wasn't her first experience with depressive episodes. They had troubled her since the death of her mother, but now, they became acute and brought on frequent panic attacks. When they came on, Marie would rush from the laboratory, convinced that the nurse had lost her baby. Once she saw that the baby was safe, she could return to work, but the stress took its toll.

Several doctors recommended she go to a sanatorium, but she refused to abandon her husband, her child, or her work. Fortunately, Pierre's father came to the rescue. He had recently lost his wife, and offered to move in with his son and daughter-in-law to help look after the household and his beloved granddaughter. With his aid, Pierre and Marie were able to continue their research.

When she recovered, Marie began working on her doctoral thesis. Wilhelm Conrad Röntgen had recently discovered x-rays. A short time later, Antoine-Henri Becquerel accidentally discovered something similar being given off by uranium salts. He shamelessly named them Becquerel rays, but then abandoned research when he found them too difficult to measure.

Marie decided to take up where Becquerel had left off. Initially, she didn't have much success measuring the strength of the rays until Pierre modified and improved his electroscope. Such devices were so notoriously difficult to work with that the scientist Lord Rayleigh once wrote all electrometers were "designed by the devil." Undeterred, Marie spent day after day laboriously measuring samples of pulverized uranium.

Once Marie had established the strength of the rays given off by pure uranium, she used that as a standard against which to measure other substances. She measured samples of the element thorium and found that they gave off rays too, although not as strongly as the uranium. She then measured a type of ore known as pitchblende. It was a heavy, black ore from which uranium was extracted. Marie found, surprisingly, that even after the extraction, the pitchblende continued to emit rays. In fact the rays she detected coming from the pitchblende were stronger than the rays from pure uranium.

Intrigued, she began exhaustively looking at other elements and compounds to determine if they gave off similar rays and if so at what strength. By April 1898 she had reached a startling conclusion. She wrote up her findings in a formal paper, but since neither she nor Pierre were members of the Academy of Sciences, her professor, Gabriel Lippmann presented the paper in her stead:

It was necessary at this point to find a new term to define this new property of matter manifested by the elements of uranium and thorium. I proposed the word radioactivity.

During the course of my research, I had occasion to examine not only simple compounds, salts and oxides, but also a great number of minerals. Certain ores containing uranium and thorium proved radioactive, but their radioactivity seemed abnormal, for it was much greater than . . . I had been led to expect. This abnormality greatly surprised us. When I had assured myself that it was not due to an error in the experiment, it became necessary to find an explanation. I then made the hypothesis that the ores of uranium and thorium contain in small quantity a substance much more strongly radioactive than either uranium or thorium itself. This substance could not be one of the known elements, because these had already been examined; it must therefore, be a new chemical element (Goldsmith, 2005).

As the scientist Frederick Soddy later put it, "Pierre Curie's greatest discovery was Marie Sklodowska. Her greatest discovery was . . . radioactivity." She had not only discovered and named an entirely new property of atoms, but had postulated that it could be used to find new elements. Shortly after Marie's paper was presented, Pierre quit his work with crystals to assist her full-time. Other scientists became interested, including Becquerel. The race to discover these new elements was on.

Soon Marie was competing against scientists in Germany, Italy, and England. She managed to isolate a substance from the pitchblende that behaved like the element bismuth, but was seventeen times more radioactive than pure uranium. As she continued to purify her samples, the radioactivity increased. Within two weeks she had produced samples 150 times as radioactive, then three hundred times as radioactive, and finally four hundred times as radioactive as pure uranium. Marie was confident that she had discovered an entirely new element, and she named it in honor of her beloved Polish homeland polonium.

She didn't stop there. By December, she found a second substance in the pitchblende that behaved like barium. This one was nine hundred times as radioactive as pure uranium. Believing she had discovered yet another new element, she wanted to confirm her results. One of her colleagues at EPCI, Eugène Demarçay was using a method called spectroscopy. It consisted of heating an element until it became a glowing gas and then using a prism to refract the light given off. The light would form a rainbow pattern or spectra, and each element gave off a unique pattern. On

December 19, 1898, Demarçay tested Marie's new sample and confirmed she had discovered a new element. She named it using the Latin word for "ray," *radius,* and christened the new element radium.

Although Marie and Pierre were certain of her results, others were unsure. The physics community was relatively comfortable with theoretical discoveries, but the chemistry community demanded something more tangible. To convince them, the Curies would need to isolate enough of the new elements to actually be measured and tested. That would prove to be a monumental task.

Pierre applied for laboratory space at the Sorbonne. He was turned down. He then applied at the EPCI, but the best they could provide was an abandoned hangar that had formerly been used by students to perform human dissections in. It had an asphalt floor and a leaky roof and was in such deplorable condition, that it was no longer suitable to house the cadavers. This was where Marie and Pierre carried out their great work.

The next hurtle was to acquire the vast amounts of pitchblende necessary to isolate a usable amount of the new elements. Fortunately, Pierre learned that there was a mountain of pitchblende sitting in a waste pile in the woods of St. Joachimsthal, then part of Austria. The uranium had already been extracted, and the piles of pitchblende sludge were simply dumped there. The Austrians were only too happy to get rid of what they considered waste material, so the Curies needed only to come up with enough money to pay for its transport. A wealthy philanthropist, Baron Edmond de Rothschild anonymously provided a modest donation for the purpose.

Once the first few tons of pitchblende arrived in the courtyard outside the hangar, the real back-breaking work could begin. It quickly became apparent that radium would be the easier of the two elements to extract, so the Curies focused on that. They divided the work, Pierre concentrating on the physics and determining its physical properties, and Marie on the chemistry of the extraction process. She carried out an exhaustive series of successive cleanings and boilings of the pitchblende in forty-four-pound batches, stirring the enormous pots with an iron rod almost as tall as she was. She then used chemical processes to purify it further and crystallize the residues. Initially, she had thought that radium made up approximately 1 percent of the pitchblende. It turned out that the actual figure was closer to one millionth of 1 percent. It took her thousands of purifications and almost seven tons of pitchblende to produce a single gram of radium.

Working tirelessly, day after day, under the most abominable of conditions, the Curies were at their best. As Marie described the time:

> We were very happy in spite of the difficult conditions under which we worked. We passed our days at the laboratory, often eating a simple student's lunch there. A great tranquility reigned in our poor, shabby hangar; occasionally, while observing an operation, we would walk up and down talking of our work, present and future. When we were cold, a cup of hot tea, drunk beside the stove, cheered us. We lived in a preoccupation as complete as that of a dream.

One evening, after tucking Irene into bed, Marie and Pierre returned to the laboratory to continue their work. As they stepped into the darkened hangar, they saw a faint glow on the shelves. It was the jars of radium salts that Marie had been purifying. Each contained only a few grains of the precious radium, but it was enough to make them glow in the dark. They reminded Marie of "faint, fairy lights," and she described her and Pierre's reaction, "From all sides we could see their slightly luminous silhouettes, and these gleaming, which seemed suspended in the darkness, stirred us with ever new emotion and enchantment" (Curie).

They were blissfully unaware that the radiation responsible for those enchanting lights was also taking a heavy toll on their health. Pierre began suffering from excruciating pain in his legs. He thought it was rheumatism and attributed it to the dampness in the hangar. His fingers became hard and cracked from handling the radium salts. Marie began to lose weight. Her doctors suspected that she had come down with tuberculosis, the disease that had killed her mother, and recommended rest and country air. She ignored them.

After four years, all of their suffering was rewarded. Marie was able to produce a tenth of a gram of radium. It was a minuscule amount, the equivalent of a few grains of sand, but it was enough for the Curies to determine its atomic weight and place it correctly as number 88 on Mendeleev's periodic table. Like Priestley before them, they shared their knowledge freely and would soon send samples of radium even to competing scientists like Ernest Rutherford, Frederick Soddy and Great Britain's Lord Kelvin.

Marie and Pierre Curie presented their discovery to the public at the 1900 World's Exposition in Paris. Just as Tesla had stolen the show with his electric demonstrations at the Chicago Exposition, center stage now went to radioactivity. A watch case containing a speck of radium was prominently displayed. Becquerel gave a talk on the history of what he called uranic rays, and then Pierre Curie presented their paper "The New Radioactive Substances and the Rays They Emit."

The first Nobel Prize for physics was given to Röntgen in 1901 for his discovery of x-rays. Marie and Pierre Curie, along with Becquerel, were nominated for the second award in 1902, but lost out to Hendrik Antoon Lorentz and Pieter Zeeman for their work on the influence of magnetism on radiation. It was a particularly disappointing loss for Pierre, because he had laid the groundwork for their research. The real disappointment, however, came the following year when Pierre Curie and Henri Becquerel were nominated for the 1903 Nobel without any mention of Marie Curie.

Almost immediately accusations started flying back and forth. It seemed to be a direct slap in the face to Marie and a complete denigration of the work of female scientists. There was also suspicion that Becquerel had helped engineer the omission as a way of boosting his own contribution. When Pierre became aware of the situation, he declared that he could not accept the prize unless it went also to Marie. Some of the committee members tried to backpedal and claim that even if they wanted to include her it was too late. The nomination had already been made. When it was pointed out to them rather forcefully that her name could be added because it had technically been included in the original 1902 nomination, they relented. Pierre and Marie Curie, along with Becquerel were awarded the 1903 Nobel Prize for physics. In hindsight, they might have been happier without it.

As part of the prize, Becquerel received seventy thousand

francs. Pierre and Marie received the same sum to split between them. They immediately used it to fund further research. In November 1903, shortly before the award ceremony, they received a formal invitation to come to Stockholm on December 10 for the award banquet. The Curies expressed their regrets and explained that it would be impossible for them to attend.

Unbeknownst to all but a few, Pierre's health had continued to deteriorate. He was now in almost constant pain and unable to sleep. What's more, that summer Marie had experienced a miscarriage. Between the toll of their exhausting work and grieving for her lost child, she was in the midst of a deep and debilitating depression. They made their apologies and promised to come at a later date.

On the night of the ceremony, Becquerel appeared alone to accept the prize. In the speeches that followed, H. R. Thornbladh, president of the Academy of Sciences credited Becquerel for his discovery of radioactivity. He gave the Curies only secondary credit for their research into the topic. In a final slight to Marie, Thornbladh acknowledged the success of "Professor and Madame Curie." Pierre was a professor, and so deserved the title, but so was Marie. By addressing her essentially as Mrs. Curie rather than Professor Curie, he had belittled her position and given her the title she would be known by in history.

As far as the Curies were concerned, the worst price they paid for the Nobel medal was the publicity it generated. Until then, the Nobel Prize was well known within the scientific community, but largely ignored by the public. That all changed when the press fell in love with the story of Madam Curie. It seemed to be the stuff of a romantic novel; the poor Polish émigré, who came to Paris and suffered alone in her cold, dark garret until Pierre, her scientist Prince Charming, lifted her to the heights of science, and together they discovered a magical, glowing substance to cure the world's ills. The public ate it up.

What followed was a storm of publicity that Pierre described as, "the disaster of our lives." They were hounded by photographers and besieged by interview requests and fan mail, even one asking for permission to name a racehorse after Marie. For two such intensely private individuals it must have been a nightmare. In a letter to a friend Pierre wrote:

> I have wanted to write to you for a long time; excuse me if I have not done so. The cause is the stupid life which I lead at present. You have seen this sudden infatuation for radium, which has resulted for us in all the advantages of a moment of popularity. We have been pursued by journalists and photographers from all countries of the world; they have gone even so far as to report the conversations between my daughter and her nurse, and to describe the black-and-white cat that lives with us. . . . Finally, the collectors of autographs, snobs, society people, and even at times, scientists, have come to see us . . . and every evening there has been a voluminous correspondence to send off. With such a state of things I feel myself invaded by a kind of stupor (Curie).

Of course, all that fame wasn't without its benefits. England's Royal Society awarded them the prestigious Davy Medal. They received numerous honorary doctorates and academy memberships from societies all over the world, and many of them were accompanied by invitations to deliver highly paid lectures. Pierre, although not Marie, was granted membership in France's Academy of Sciences. The French Parliament, the Chamber of Deputies, created a special science chair for Pierre at the Sorbonne. It came with a salary of ten thousand francs, and he managed to parlay his influence to have them include a fully equipped laboratory with three paid assistants and a position for Marie as head of research.

As fascinated as the public was by the Curies, it was absolutely mesmerized by the miraculous substance radium. The new element was touted as a cure for everything from cancer to infertility. All manner of quacks and snake oil salesmen started using *radium* and *radioactive* as buzz words in their advertisements. Pure radium was much too precious for most such uses, but it was often diluted, sometimes by as much as six hundred thousand times, and mixed with zinc and other substances. These adulterated mixtures were then added to tonics, teas, bath salts, even cosmetics. George Bernard Shaw wrote, "The world has run raving mad on the subject of radium, which has excited our credulity precisely as the apparitions at Lourdes excited the credulity of Roman Catholics."

To escape the circus-like atmosphere being generated around them, the Curies left Paris for a summer vacation on the Normandy coast. They rented a small cottage and invited Marie's sister Helena and her young daughter, Hania, to join them. Although Pierre's health continued to deteriorate, Marie was able to enjoy herself, swimming and helping the girls collect shells. Marie became pregnant, and vowed to take better care of herself this time. On December 6, 1904, she gave birth to a healthy daughter, who she and Pierre named Eve.

By the summer of 1905, things had calmed a bit, and the Curies were at last able to travel to Stockholm and deliver the lecture customary for Nobel recipients. In a further slight to Marie, only Pierre was asked to speak. Those who wished to rob her of her accomplishment, however, were disappointed. As Pierre took the podium, with Marie seated in the audience, he delivered an address that gave her the full credit she was due. He noted his own contributions, but again and again pointed out that Marie Curie had discovered radioactivity and had proven the existence of both polonium and radium.

The following spring, Pierre and Marie had another chance to take a brief holiday with their children. They traveled to St. Remy-les-Chevreuse in northern France for the Easter break. Sitting side by side, while Irene chased butterflies and fourteen-month-old Eve lay in the sunshine on her blanket, Marie and Pierre had the luxury of enjoying each other's company. Unfortunately, the demands of work intruded once again, and Pierre was forced to go back a day or two before Marie and the children. On Monday, he returned to Paris with a bouquet of wildflowers that he and Marie had picked.

Marie and the girls came home on Wednesday. On the Thursday morning of April 19, as Pierre rushed to leave for the lab, they had a brief argument. He wanted Marie to go to the lab with him, but she wanted to take Irene on an outing. As he demanded to know if she was going to meet him at the laboratory, Marie snapped back, "I don't know. . . . Don't torment me." With that, Pierre took up his umbrella and stepped out into the drenching spring rain.

He spent the morning working and then attended a luncheon meeting of the Association of Professors of the Science Faculties. When lunch concluded, Pierre headed out once more into the rain. His path took him to the intersection of the Pont Neuf and the rue

Dauphine, one of the busiest in Paris. That afternoon, the sidewalks were jammed and the streets busy with wagons, carriages, cabs, and busses. His view partially obscured by his umbrella, Pierre stepped off of the rain-slick curb and into the street. Just then, a heavily loaded wagon, pulled by a team of Percherons came by. Pierre tried to avoid them, but his weakened legs betrayed him. He stumbled and fell beneath the wheels dying instantly as one of the rear wheels crushed his skull.

That evening, Marie and Irene returned from their outing and found Paul Appell, a long-time family friend and dean of the department at the Sorbonne, waiting for them. He told Marie about what happened, and when he was done, she mumbled incredulously, "Pierre is dead. Dead. Absolutely dead?" As she tried to make arrangements for the body, Marie noticed the bouquet of wildflowers that Pierre had put in a vase of water. She simply walked out into the rain-soaked garden and sat on a bench, "elbows on her knees and her head in her hands, her gaze empty. Deaf, inert, mute" (E. Curie, 1937).

At his funeral, Marie tried to give her Pierre, her love, her soul mate, her partner, one final respite from the maddening crowds. She insisted that the service be a simple one without speeches or intrusions by the public. She was unaware that there were reporters hiding in the cemetery.

To friends and even close family, Marie presented a stoic facade, bravely bearing her loss. In private she began a journal in which she poured out her heart:

On the Sunday morning after your death, Pierre, I went to the laboratory with Jacques for the first time. I tried to make a measurement, for a graph on which we had each made several points. But I felt the impossibility of going on.

In the street I walk as if hypnotized, without attending to anything. I shall not kill myself. I have not even the desire for suicide. But among all these vehicles is there not one to make me share the fate of my beloved? (M. Curie, 1923).

If she had been tempted to give up their dream, Marie only had to remember Pierre's own words. When his illness became undeniable, and the two of them had been forced to confront his mortality, Marie confessed to Pierre that without him, she would be unable to work. He gently reprimanded her and told Marie, "It is necessary to continue no matter what."

The French government offered her a pension, but Marie refused. She continued her work. Officials at the Sorbonne were at a loss as to how to fill the void created by Pierre Curie's death. Who was qualified to replace him in the laboratory and at his lectures? There was only one choice. It would take the officials two years to formally acknowledge it, but when the fall term started Marie Curie effectively filled the chair created for her husband, the first woman in the history of the Sorbonne to hold such a position.

Her first lecture was scheduled for 1:30 in the afternoon of November 5, 1906. The lecture hall officially had a capacity of 120, but by 10:00 that morning, hundreds of people had lined up outside. When the doors opened at 1:15, a veritable flood of reporters, photographers, academics, students, Parisian high society, and members of the curious public rushed in. In the pandemonium that followed, a small, black-clad figure slipped in through one of the back doors. As she took her place behind the bench where her husband had once lectured, applause erupted for Marie Curie.

Once the crowd had calmed down, she began with quiet dignity:

When we examine our recent progress in the domain of physics, a period of time that comprises only a dozen years, we are certainly struck by an evolution that has nourished fundamental notions regarding the nature of electricity and of matter. This evolution happened in part because of detailed research on the electrical conductibility of gases, and also because of the discovery and study of the phenomena of radioactivity.

She had begun her lecture, exactly where Pierre had left off. When she was done, the crowd rose as one, and gave her a standing ovation. The next day, she wrote in her journal, "Yesterday I gave the first class replacing my Pierre. What grief and what despair! You would have been happy to see me as a professor at the Sorbonne, and I myself would have so willingly done it for you, but to do it in your place, my Pierre, could one dream of a thing more cruel?"

The work continued in the laboratory as well. Under Marie's direction, the lab investigated medical and industrial uses for radium. With the greater resources now at her disposal, Marie could begin the process of isolating polonium. It turned out that the concentration of polonium in the pitchblende was four thousand times less than that of radium, but after four years of work, in 1910, she succeeded. By then, the Curie laboratory, with

Marie's leadership, grew from six employees to twenty-two. In addition, it employed twenty women scientists who volunteered their time.

Because of these successes, there were those who felt threatened by Madam Curie. They accused her of being simply an assistant who had ridden to fame on her husband's coattails. With the encouragement of friends, Marie applied for membership in the French Academy of Science. Many supported her, but there were also members who were outraged at the thought of letting a woman into their ranks. When the vote was taken in 1911, her proposed membership was defeated by one vote. It would be sixty-eight years before the first woman was admitted to the academy.

Shortly after her defeat, rumors began circulating that Marie Curie was having an affair with a married man. The man in question was fellow scientist, Paul Langevin. He had been one of Pierre's students and had succeeded him at the EPCI. Now a brilliant researcher in his own right, Langevin was known to have been in a tempestuous and reportedly unhappy marriage. His wife had tolerated his previous infidelities, but when she learned of the affair with the famous Madam Curie, she became enraged. His wife hired a detective who stole a series of letters between Marie and Langevin, and she threatened to turn them over to the press.

When the controversy was at its height, the public that had previously adored Marie Curie now turned on her. She was denounced in the press, and in the streets met with calls of, "home wrecker" and "husband snatcher." A group of professors at the Sorbonne, including her former friend Paul Appell, who had brought her news of Pierre's death, demanded that she leave

France. It is interesting that no such demands were placed on Langevin.

Ironically, in the midst of this chaos, while Marie and Langevin were both at a scientific conference in Belgium, she received a telegram informing her that she had won a second Nobel Prize, this time in chemistry, for her discovery of radium and polonium. Marie Curie had become the first person in history, male or female, to win two Nobel Prizes.

Shortly after that, however, she received a second telegram telling her that Langevin's wife had been good to her threat, and had released the letters. Within days, a private letter from one of the members of the Nobel Committee arrived asking her to refrain from coming to Stockholm to accept her award. It included the following, "If the Academy had believed the letters . . . might be authentic it would not, in all probability, have given you the Prize."

Rather than demurely acquiescing, Marie met the controversy head on. In her response she wrote:

> You suggest to me . . . that the Academy of Stockholm, if it had been forewarned, would probably have decided not to give me the Prize, unless I could publicly explain the attacks of which I have been the object. . . . I must therefore act according to my convictions. . . . The action that you advise would appear to be a grave error on my part. In fact the Prize has been awarded for discovery of Radium and Polonium. I believe that there is no connection between my scientific work and the facts of private life. . . . I cannot accept the idea in principle that the appreciation of the value of scientific work should be influenced by libel and slander concerning

private life. I am convinced that this opinion is shared by many people.

Marie Curie went to Stockholm, received her prize, gave her acceptance speech, and nothing more was said about personal matters. Eventually, Paul Langevin was granted a separation from his wife. Years later, they reconciled, and he continued to take other lovers. The controversy faded away as the ever fickle public found other things to worry about.

By the summer of 1914, all French eyes were on Germany. On August 3, war was declared and German troops began marching toward Paris. The newly completed Radium Institute, whose construction Marie had overseen, now lay all but deserted, as was most of the city. As artillery shells rained down on Paris, Marie Curie remained to protect her precious radium and ensure that it didn't fall into enemy hands. When she finished loading the lead-encased tubes of radium bromide into her suitcase, she lugged the heavy bag to the rail station and boarded a train for Bordeaux. Once there, she saw her burden deposited safely in a bank vault and returned once more to Paris.

As the horribly wounded began returning from the front, it became apparent that their wounds were made worse by surgeons trying to remove bullets and shrapnel without benefit of x-rays. Marie at once went to work commandeering the needed equipment and setting it up in the military hospitals. Then she had an inspiration. Instead of transporting the wounded long distances to the x-ray equipment, why not take the equipment to them? She procured a few cars and began creating the first mobile x-ray units. When the first units rolled out, Marie Curie herself was behind the wheel, dressed in an alpaca coat with a Red

Cross armband on her sleeve. These lifesaving units came to be known as "Les Petites Curie."

Just before the war, Marie Curie had been instrumental in establishing an international standard for measuring radiation. Her laboratory produced the radium against which all other radioactive sources would be measured. In 1913, she hand delivered a glass tube containing 21.99 milligrams of radium chloride to the Bureau of Weights and Standards in Sevres, France. It should be noted that she retained a duplicate sample for safekeeping at her institute. In recognition of her contribution, the Standard Committee named the new unit used to measure radiation the curie.

In later years, Marie, like her beloved Pierre, experienced a series of serious health problems. She was constantly fatigued and suffered from severe tinnitus or ringing in the ears. In spite of operations for cataracts, she was nearly blind. Even though her doctors were looking for ways to diagnose her illnesses, she stubbornly refused to take the blood tests that she required of her workers.

Marie's condition continued to deteriorate. On July 3, 1934, Marie Curie lapsed into a coma and died. The cause of death was listed as "aplastic pernicious anemia." In other words, her bone marrow had been so damaged by years of radiation exposure that it was unable to make enough red blood cells. The doctor remarked that it was a miracle that she had lived to the age of sixty-seven. Marie was buried next to Pierre in Sceaux.

Marie Curie became a legend, her genius and single-minded determination in the face of fierce resistance glowed as brightly as the radium that she discovered. She once described radium, the substance that she had dedicated so much of her life investigating as "my child." How ironic that like the fictional Victor Franken-stein, the scientific child that she helped usher into the world would lead to her demise. For those, however, who think of her story as tragic, they should remember her own words:

> I am among those who think that science has great beauty. A scientist in his laboratory is not only a technician, he is also a child placed before natural phenomena, which impress him like a fairy tale. We should not allow it to be believed that all scientific progresses can be reduced to mechanism. . . . Neither do I believe that the spirit of adventure runs any risk of disappearing in our world. If I see anything vital around me, it is precisely that spirit of adventure, which seems inde-structible (E. Curie, 1937).

A wealthy Belgian chemist and industrialist, Ernest Solvay de-cided to invite the world's greatest physicist to a conference in 1911. It was the first of what would become known as the Solvay Conference. The invitation-only event was the first worldwide meeting of physicists and would go on to influence the shape of modern physics. Only the finest physicists in the world were invited. Marie Curie was naturally among this select group of scientific luminaries. While there, she met a brash, young physi-cist, a German Jew and the youngest scientist at the conference. His unconventional ideas were beginning to create quite a bit of buzz in the physics community.

His theories would go on to ignite a scientific revolution that toppled those of the great Isaac Newton himself. As his fame spread across the planet, he would become one of the best-known and most influential figures of the twentieth century. His name would come to be synonymous with the word *genius,* and he, perhaps more than any other figure, would shape the public's image of the mad scientist. His name was Albert Einstein.

The medieval German city of Ulm lies where the Blau and Iller Rivers flow into the Danube. There on March 14, 1879, Albert Einstein was born. His father, Hermann, was a middle-class businessman, a partner in a featherbed company. Like Semmelweis' father, he was of Swabian decent. Albert's mother came from a wealthy family of grain merchants. Although both parents were Jewish, neither was particularly observant, and Albert was, for all intents and purposes, raised in a secular household.

As the story goes, young Albert was slow learning to talk. So much so, that his parents consulted doctors. Even when he did begin speaking, at around the age of two, he developed a strange habit of whispering to himself before he said anything, as if trying out each word before he dared say it out loud. His parents continued to worry, but when Albert was five, something happened that would start him down the path to becoming one of the greatest scientists in history. His father gave him a compass.

Albert was sick in bed and his father, wanting to cheer the boy up, gave him a small magnetic compass to play with. Once Albert noticed how the compass' needle stubbornly pointed north, no matter which direction he turned it, the boy became entranced by the magic of the invisible force controlling the needle. As he recounted it years later, "I remember—or at least I believe I can remember—that this experience made a deep and lasting impression on me." He went on to say, "Something deeply hidden had to be behind things" (Isaacson, 2007).

Hermann moved the family to Bavaria, so he could join his

brother Jakob's electrical engineering firm. Albert began school there in Munich. Another commonly told story says that Einstein failed math. It's an amusing story, but it doesn't seem to have much basis in fact. In primary school, he was at the top of his class. By the age of twelve, he was learning geometry and algebra. His uncle Jakob, the engineer, would often challenge the boy with math problems and puzzles, which young Albert invariably solved.

Albert showed great interest in science and mathematics, but this didn't translate into a general interest in school. He was a rebellious child, who constantly questioned authority. His younger sister Maria described it this way, "The military tone of the school, the systematic training in the worship of authority that was supposed to accustom pupils at an early age to military discipline, was particularly unpleasant" (Isaacson, 2007). As the other children played soldier and happily joined in the frequent military parades, Albert looked at them with a disdain that would later be expressed in his pacifist views. As he said, "When a person can take pleasure in marching in step to a piece of music it is enough to make me despise him. He has been given his big brain only by mistake" (Einstein, 1930).

By the time Albert was fifteen, his father and uncle's engineering company was in serious trouble. They had lost a number of important contracts, and by 1894 the company was bankrupt. Most of the family moved to the northern Italian city of Milan hoping for better fortune, but Albert was sent to live with relatives in Munich so he could finish his education. Albert had other plans. He dropped out of school and followed his family to Milan. The headstrong youth informed his family he intended to complete his studies on his own and try to attend a technical college in Zurich. His decision was likely influenced by the fact that, had he remained in Germany until the age of seventeen, he would have been subject to military conscription. With that in mind, he also renounced his German citizenship.

By 1896, the seventeen-year-old Einstein was enrolled in the Zurich Polytechnic. During the years of self-education, he had become quite interested in physics and chose that as his major. His physics professor was Heinrich Weber, whom he initially got along well with. Eventually, however, Einstein found his professor too conservative and resistant to exploring the newer views emerging in the field. Einstein's rebellious nature once again asserted itself. One day Weber told him, "You're a very clever boy, Einstein, an extremely clever boy. But you have one great fault: you'll never let yourself be told anything" (Isaacson, 2007).

That attitude would serve him well as an innovative young physicist, but as a student, it caused problems. He took a course titled "Physical Experiments for Beginners," taught by professor Jean Pernet. Einstein, feeling insufficiently challenged, frequently skipped the class. When he did attend, and it came time for labs, young Einstein would often throw out the instructions and begin experimenting on his own. Eventually, this led to an accident, causing an explosion in the laboratory. Einstein severely injured his right hand in the blast, and from then on showed a decided propensity for theoretical rather than experimental physics. For Pernet's part, he can go down in history as the professor who failed Einstein in physics.

When not antagonizing his professors, Einstein made the acquaintance of Mileva Maric, the only female physics student in his class. It wasn't exactly love at first sight. She came from a Serbian family, and was short, dark, and brooding. Einstein, on the other hand, was considered quite the ladies' man, with his sparkling eyes and quick wit. The two, however, formed an intense and passionate relationship, driven as much by their ability to challenge each other intellectually as by romance. Their love letters were filled equally with flirtatious banter and critiques of the latest physics theories.

By the time Einstein graduated in 1900, the pair had become lovers, and by the following spring, Mileva was pregnant. Both of their families were horrified. The situation was made worse because Einstein, the expectant father, was unemployed. He applied to be a professor's assistant at the polytechnic, but had so alienated most of his professors that they refused to give him a job. He tried applying for other positions at other universities with similar results. Even when he had a minor paper published in one of Europe's leading physics journals, they still refused to hire him. He was forced to make do with minor tutoring jobs.

While Einstein struggled to find meaningful employment, Mileva endured her own struggles. She failed her final exams and was unable to graduate. As her pregnancy grew more and more apparent, it became too scandalous for her to be seen with Einstein in public. She ended up spending much of the later part of pregnancy in a small hotel in a nearby village. At last, in February 1902, after a difficult delivery, Mileva gave birth. They named the baby girl Lieserl. When she was born, Einstein was off looking for work in Bern.

Einstein got a lead on a job in the Swiss patent office. The previous year, he had obtained Swiss citizenship, so was qualified to be a civil servant. It was a modest position, with the title technical expert class 3, but it was something. With a regular salary in hand, he got an apartment for himself and Mileva and anxiously wrote to her of his new position. Unfortunately, a proper Swiss civil servant could not be seen to have an illegitimate child, and Einstein would never actually lay eyes on his daughter.

There is a great deal of confusion about what actually happened to the girl. Mileva was staying with her family when Lieserl was born. She wrote to Einstein frequently with news of the baby, but mysteriously in September 1903, all mention of the girl stops. There's some evidence that Lieserl came down with scarlet fever, causing speculation that she died, but there are no records of her death. There are other hints that Lieserl may have been adopted by one of Mileva's close friends. Whatever the cause, no further mention of the daughter is made.

Back in Bern, Albert Einstein adapted to the role of a third-class patent clerk. Six days a week, he walked to work past the city's famous clock tower, and eight hours a day evaluated patent applications. He seemed quite happy with the job, and his boss was happy with his work. Once he was established, Mileva moved to Bern, and on January 6, 1903, they were married. It was a civil service, and neither family was present, but their friends helped them celebrate. Afterward, they returned to the apartment and had to wake the landlady. Einstein, already displaying the absentmindedness that would later become famous, had forgotten the key.

If 1666 is often celebrated as Isaac Newton's *annus mirabilis* (miracle year), then 1905 was Einstein's. One of the things he liked most about his new job was that he was able to get the actual work done quickly, leaving him plenty of time to think. His boss seemed not to mind the piles of scratch paper full of obtuse calculations that filled his desk, and Einstein would spend many hours conducting thought experiments. These led him to produce not one, but four extraordinary physics papers, any one of which could have been career making for most physicists.

The first dealt with the nature of light. Scientists, including Newton, had been arguing for centuries over whether light was composed of tiny particles or whether it was a wave. Most of the evidence seemed to be leaning toward the wave theory, but there were a few curious anomalies. The first had to do with something called the photoelectric effect, the way that metals give off electrons when bombarded with light. Experimenter Philipp Lenard discovered when he increased the frequency of the light, from infrared up to ultraviolet, the electrons were emitted with greater energy, but when he increased the intensity of the light, their energy remained the same. There were simply more of them emitted.

The second anomaly had to do with the work of German physicist, Max Planck. He was working with the way objects give off light when heated, called black body radiation. In the course of his calculations, Planck came across a mathematical constant (approximately 6.62607×10^{-34}) that was necessary for the calculations to come out correctly. He wasn't quite sure what this was, but proved that it was needed to explain the effects of the heat. It came to be known as Planck's constant.

Einstein used both anomalies and did the math needed to show that light wasn't exclusively a wave or a particle but rather something more subtle, called quanta. These were discreet packets of energy simultaneously able to act as both particles and waves. They eventually came to be known as photons. This explained both the photoelectric effect and the behavior of black body radiation. Together with Planck's work, it lay the groundwork for the entire field of quantum mechanics.

Einstein's second revolutionary paper was actually his doctoral dissertation. In the 1800s an Italian scientist named Amedeo Avogadro was working on the properties of gases when he discovered the number of particles needed to form a mole of a gas (its molecular weight in grams) was approximately 6.02214×10^{23}. This, as any high school chemistry student can tell you, came to be known as Avogadro's number. Einstein applied this to liquids and proposed a way to calculate it experimentally using the liquid's viscosity, the resistance to objects moving through the liquid. This had many practical applications in fields ranging from mixing concrete to the production of dairy products. It also got Einstein his PhD. He was now Dr. Einstein. Of course, this still didn't help him get an academic job, but it did earn him a promotion from third class patent clerk to second class.

His third paper, which he finished eleven days after his dissertation, was on Brownian motion. When small particles are suspended in a liquid, they appear to be in constant random motion. The phenomenon was named after the scientist who first documented it, Robert Brown. Many possible explanations had been given, but none had been proven. Einstein created a mathematical model to show that the motion was caused by the cumulative effects of the millions of collisions occurring between the suspended particles and the liquid's atoms.

He then used the model to make specific predictions about the motion and the size of atoms, and suggested experiments that could be used to test his predictions. A few months later, a German experimenter named Henry Seidentopf proved experimentally that Einstein was right. Since the time of the ancient Greek philosopher Democritus, people had been arguing over the existence of atoms. Einstein's paper gave actual proof of their existence and laid the argument to rest.

Einstein's crowning achievement of 1905 was his paper proposing a special theory of relativity. He began with the assumption that all motion was relative. In other words, a person sitting in his living room will have the same experience of motion as a person sitting in an airplane flying overhead at hundreds of miles per hour. Neither of them will notice any difference as they reach for their drink. The ice cubes in the drink in the living room will behave exactly as the ice cubes in the drink on the plane. Each person can talk to the person next to them without noticing any increase in the speed of the sound. If they accidentally drop something, they will both see it fall straight to the floor. Each can comfortably assume that they are the one standing still and the other person is the one rushing past at hundreds of miles per hour. For them, all motion is relative.

When he was a boy, Einstein had daydreamed about what it would be like to ride along next to a beam of light. Now he took that thought experiment and expanded on it. He imagined himself racing the light beam as it sped away from Bern's famous clock tower. Then he thought about what he would see as he looked back at the clock. As he raced faster and faster through space, farther and farther from the clock, he could see in his mind's eye that the hands of the clock began to slow down as the light from them tried to keep up with him.

Eventually, as he achieved the speed of light, the hands stopped. Time stood still. As the realization hit him, Einstein described, "A storm broke loose in my mind" (Kaku, 2004). Until that moment, scientists had conceived of absolute space and absolute time as separate things. What Einstein did was fuse them together into an inseparable whole he called space/time.

In the course of doing his calculations on light and relative motion, almost as an afterthought, Einstein came up with an interesting equation, $E = mc^2$. E represented energy. The letter m represented mass, and c represented a constant equal to the speed of light. The speed of light is a big number, 186,000 miles per second. If you take a big number and square it, you get an unbelievably big number. Essentially then, this equation says that a quantity of mass times an unbelievably big number will equal an immensely large amount of energy. Instead of being two separate things, matter and energy are essentially two facets of the same thing, and you can convert one into the other. This not only knocked the familiar laws of conservation of mass and energy for a loop but would also have explosive implications in a few years to come.

With youthful self-assurance, Einstein sent his papers off into the world to be published, secure in the knowledge that success and academic acclaim would soon be his. He simply waited for a reply, and waited, and waited. After months, he heard nothing. Once again the hands of the clock seemed to slow down and stop as Einstein waited what must have seemed an eternity for the recognition he deserved. Fortunately for the young patent clerk, his papers eventually landed on the desk of perhaps the one person fully capable of appreciating their implications, the most important theoretical physicist in the world, Max Planck.

Planck, in addition to working with black body radiation, was editor of the journal *Annalen der Physik* (Annals of Physics), and had been responsible for vetting Einstein's paper. He was intrigued by this new theory of special relativity. In fact, shortly after it was published, Planck gave a lecture on Einstein's theory, and published his own article arguing in favor of it. Soon, Einstein was exchanging letters with Planck, who was still unaware that the author of this revolutionary theory was an unknown clerk.

Planck sent his assistant, Max von Laue to Bern to meet the young physicist. It came as quite a surprise to von Laue when he learned that Einstein was not at the University of Bern, as he had assumed, but working downtown in a government office. As he sat in the patent office reception area, he recalled, "The young man who came to meet me made so unexpected an impression on me that I did not believe he could possibly be the father of the relativity theory, so I let him pass."

After that initial awkward start, von Laue and Einstein carried out a series of long conversations and correspondence, eventually becoming close friends, but acceptance of Einstein's radical theory didn't come quickly or easily to everyone. Many were uncomfortable with its abstract nature. Some even suggested that the entire thing was just a little too Jewish. Arnold Sommerfeld, who later became a friend of Einstein's, said in a letter to a colleague, "As remarkable as Einstein's papers are, it still seems to me that something almost unhealthy lies in this unconstruable and impossible to visualize dogma. An Englishman would hardly have given us this theory. It might be here too, as in the case of Cohn, the abstract conceptual character of the Semite expresses itself" (Hoffmann, 1972).

Despite all this attention, Einstein was still unable to get a job in academia. He continued to work in the patent office, and he and Mileva settled into domestic life. They even had a second child, a boy named Hans Albert. Einstein continued to submit papers. By 1907, however, while working on an article about his theory of special relativity for a science yearbook, Einstein started to become troubled. It seemed to him that his theory was incomplete, that it had two yawning holes, and each must have looked to Einstein like it was a light-year across. The first was that the theory applied only to objects that were in constant uniform motion. In the real world, objects are usually accelerating, speeding up, or slowing down. Special relativity didn't account for this. In fact, that's why he called it *special* relativity, because it only applied to that one special case.

The second hole was that it failed to take into account gravity. Einstein wanted to create a more general theory that could include gravity. His breakthrough came while he was at work, as usual staring out the window. There were a number of tall buildings across the street, and Einstein began wondering what would happen if one of the workmen on the buildings fell off the roof. As Einstein visualized the hapless workman falling, he realized that if he were encased in a small room, like an elevator he wouldn't see himself falling. Rather, he would see himself and everything else in the elevator begin to float, because he and the elevator would be falling at exactly the same rate.

This is familiar to anyone who has seen footage of astronauts floating in the space shuttle. The astronauts and the shuttle are actually falling around the Earth, so they appear to be weightless. Einstein, however, had never seen a space shuttle. He used his thought experiment of the plummeting man in the elevator to begin generalizing his theory of relativity to include gravity.

It was a direct challenge to Newton's views, and when Einstein told his friend Planck about what he was doing, Planck warned him, "As an older friend, I must advise you against it for in the first place you will not succeed, and even if you succeed, no one will believe you." Of course, Planck also went on to tell his young friend, "If you are successful, you will be called the next Copernicus." How could Einstein resist a challenge like that?

He was, however, still stuck at the patent office. Even while consulting with some of the most prominent physicists in Europe, he was still unable to get a teaching position. He went so far as to apply for a position as a high school mathematics instructor. The job went to someone else. Eventually, in 1908, he managed to get a position as a Privatdozent at the University of Bern, similar to the position that Semmelweis had been forced to take. It was unpaid, so Einstein had to keep his patent office job, but it allowed him to teach, and at least it was a foot in the academic door.

It was a smart move on Einstein's part, because later that year the University of Zurich decided to create an associate professorship in theoretical physics. Einstein was the obvious man for the job, but initially the university wanted to give the position to Friedrich Adler, one of his friends. He and Einstein had been classmates at the polytechnic. Adler had a number of political contacts, because his father was head of the Social Democratic Party in Austria, but when Adler found out that Einstein might be denied the job, he was outraged. He wrote to his father and told him, "The scandal is being felt not only here but in Germany that such a man would have to sit in a patent office" (Isaacson, 2007). Shortly after that, Einstein was given the job. He was now a professor, or as he told a colleague, "So, now I too am an official member of the guild of whores."

When the announcement of Einstein's appointment appeared

in the paper, one of his ex-girlfriends wrote to congratulate him. In his response, Einstein mentioned that he and Mileva were married, but told the woman that if she were ever in Zurich to drop by. Unfortunately for everyone concerned, her response was intercepted by Mileva, who flew into a jealous rage. With some difficulty, Einstein managed to smooth things over, but his relationship with Mileva would never be the same.

Mileva became increasingly jealous, not only of Einstein's possible romantic entanglements but also of his professional ones. As his success grew, he spent more and more time in work, away from his family. Mileva, who once aspired to a scientific career herself, was now relegated to the role of house frau. Even as she helped him with his work, checking his math, typing his manuscripts, and acting as a sounding board for his ideas, Mileva became increasingly embittered. Absorbed in his work, Einstein was largely oblivious.

While his home life deteriorated, Einstein began butting heads with Max Planck as well. The two could not have been more different, Planck the impeccably dressed, always formal, autocratic Prussian, and Einstein, the slovenly young Bohemian. Although Planck had helped usher in the revolution in physics, he was uncomfortable with its implications. Increasingly, he was put in the role of defender of the orthodox, while Einstein was cast as the revolutionary Young Turk. Despite this, both men continued to have great mutual respect for each other.

Their debate was given a very public stage at the Solvay Conference in 1911. As mentioned earlier, this was the first world conference of physics. Only the most prominent physicists were invited, including Max Planck, Henri Poincare, Ernest Rutherford, Hendrik Lorentz, Paul Langevin, and Marie Currie. Einstein, at the age of thirty-two, was the youngest of the group, and one of only eight asked to present a paper. This allowed him a forum in which to defend his theories.

Unfortunately, much of the discussion was sidelined by the furor over the Curie/Langevin affair. Einstein, who had just met Curie was disgusted at her treatment by the press, and wrote her a note advising, "If the rabble continues to occupy itself with you, then simply don't read that hogwash, but rather leave it to the reptile for whom it has been fabricated" (Isaacson, 2007).

After the conference, Einstein returned home, where things were not nearly as pleasant. He and Mileva were living in Prague where Einstein had accepted a prestigious position as a full professor. A year earlier, they had another child, Eduard. It had been a difficult birth, and Mileva was sick for weeks afterward. Even when she recovered physically, she became depressed, and complained frequently about how she hated Prague. Her husband's increasing success was an ever-present reminder of the career she had been denied. Some of their friends suspected that she might be suffering from schizophrenia. Rather than facing the problems at home, Einstein retreated once more into his work.

Frequently, Einstein's work required him to travel, and in the spring of 1912 he went to Berlin. Mileva remained home with the children. While in Berlin, Einstein became reacquainted with his cousin Elsa. They were first cousins and had played together frequently as children. Now Elsa was divorced and living with her two daughters. To Einstein, she must have seemed a pleasant change from his wife. While Mileva was dark, intellectual, and moody, Elsa was open and nurturing. The two fell in love, and even when he returned to Prague, Einstein and Elsa maintained a secret correspondence.

In a last-ditch attempt to save his marriage, Einstein eventually told Elsa that he couldn't write to her any longer and tried to get a position at his old alma mater in Zurich, the polytechnic. He knew that Mileva loved the city, and they had many friends there. The polytechnic had since become a full university and been renamed the Swiss Federal Institute of Technology, and Einstein should have had no trouble getting a position there, but he met resistance. Some officials tried making excuses; they couldn't afford a professor of theoretical physics, or they didn't have the lab space. It was pointed out to them that in this day and age, theoretical physics was a necessity not a luxury, and Einstein needed no more laboratory than a blackboard and a desk. Amusingly, the university received letters of recommendation on Einstein's behalf written by Marie Curie and Henri Poincare, two of the best-known scientists in the world. Shortly after that he was given the job.

Meanwhile, Max Planck had his own designs on the rising physics star. Planck and his chemist colleague, Walther Nernst, had been charged with creating a one-of-a-kind scientific institute in Berlin. It was to host the finest scientific minds in the world, and establish Berlin as the scientific center of cutting-edge research. Planck knew that Einstein would generate the excitement and publicity that such a place needed.

In July 1913, Planck and Nernst boarded a train for Zurich intent on securing their prize. When Einstein greeted them at the train station, they presented him with an incredible offer. They offered Einstein a position as head of the Kaiser Wilhelm Institute of Physics. The job came with a handsome salary and all the resources and assistants he could ever need. It also required almost no teaching duties, leaving him free to do research full time.

What's more, he would be inducted into the prestigious Prussian Institute of Science, the youngest member in its history. Of course, the position required that he renew his German citizenship, but they assured Einstein that he could keep his Swiss citizenship as well.

Even with such a magnanimous offer before him, Einstein characteristically told Planck and Nernst that he needed time to think. Perhaps he was hesitant or perhaps he simply enjoyed keeping them on pins and needles for a bit, but for whatever reason, Einstein told the two to spend some time seeing the town, and that afternoon, when it was time for them to take the return train, he would give them his answer. With a lighthearted bit of panache, Einstein told Planck and Nernst that he would meet them on the train platform with a bunch of flowers. If the flowers were white, it meant he was turning them down. If the flowers where red, it meant he would be coming to Berlin. They parted, and that afternoon, when the two German recruiters returned, there was Einstein on the platform, holding a bunch of red flowers.

Mileva was less than thrilled. She loved Zurich and thought of it as home. Einstein had a secure position and a lovely home. Mileva knew that as a Slav, she would not be terribly welcome in the Teutonic society circles of Berlin, and to make matters worse, living in Berlin would mean living closer to her mother-in-law. The two did not get along well. Einstein, on the other hand, was thrilled by the new challenge and told his friends, "I could not resist the temptation to accept a position in which I am relieved of all responsibilities so that I can give myself over completely to rumination" (Isaacson, 2007). Of course, that was not Berlin's only temptation. It was also home to Elsa.

The Einsteins moved to Berlin in the spring of 1914. It was

effectively the death knell for their marriage. Einstein struggled to adjust to his new job, while still wrestling to generalize his theory of relativity. He spent more and more time away from his family. In his absence, Mileva's mood continued to darken. In response to troubles at home, Einstein increasingly retreated into his work. It was a downward spiral.

By July, it was over. Einstein and Mileva reached a separation agreement. Mileva got custody of the children, and Einstein provided them with fifty-six hundred marks per year, about half his salary. On July 29, 1914, Mileva and the two boys boarded a train for Zurich. One of Einstein's friends who accompanied him to the station described how Einstein, a man who took pride in his detachment from personal matters "bawled like a little boy" all afternoon and into the night.

Einstein continued to work on his general theory of relativity. He was trying to expand on his original theory to have it take into account acceleration, specifically the acceleration caused by gravity. Newton had explained that gravity was a force that made objects accelerate toward each other and had worked out elegant mathematical calculations to describe how fast they accelerated and how gravity could hold the planets in orbit. Unfortunately, he neglected to explain why gravity made things accelerate. He simply took it for granted that it did.

Einstein began once again contemplating his imaginary workman in the falling elevator. If the elevator weren't falling, if it were standing still, the workman would feel the effects of gravity. He would feel his feet planted firmly against the floor of the elevator. If he dropped some change from his pocket, he would see it fall to the floor. Now imagine that the elevator was in deep space. Imagine that it's tethered to a rocket moving perpendicular to the elevator's floor. As the rocket makes the elevator accelerate, the workman would again feel his feet planted against the floor, this time not by gravity, but by the acceleration. If he dropped more change, it would once again fall to the floor. In other words, he wouldn't be able to tell the difference between the elevator sitting on Earth, and the elevator accelerating in deep space. Einstein's great breakthrough came in realizing that if the workman couldn't tell the difference, then perhaps there wasn't one. Perhaps acceleration and gravity were one and the same.

Now let's assume that the accelerating elevator has a small window, and a beam of light is coming through the window shining on the opposite wall. If the elevator were accelerating fast enough, then the beam of light wouldn't hit exactly opposite the window but would hit a spot on the opposite wall lower than the window. Essentially, the spot directly opposite the window would have accelerated out of the way, causing the beam of light to miss. To the workman in the elevator, it would look as if the beam of light bent. Now if acceleration bends the light, and acceleration and gravity are indistinguishable, then that means that gravity will bend the light.

Of course, everyone knows that light doesn't bend. It travels in a straight line through space. Unless, that is, space is bent. If gravity bends space, then it would bend the light traveling through space. It would also explain why objects accelerate because of gravity. The gravity is bending the space the object moves through. Think about a trampoline. Now imagine that you put a bowling ball in the center of it. That causes the surface of the trampoline to bend. If you roll a marble across that surface, instead of going in a straight line, it will be drawn into the depression caused by the bowling ball. If it doesn't have enough momentum to escape, it'll begin orbiting around the bowling ball, just like a planet orbiting around the sun.

Einstein was on the verge of creating what would one day be called one of the most elegant theories in the history of science. Of course, now he needed proof. If his theory were correct, then the light from distant stars would be bent as it passed near a sufficiently large gravitational mass. That would cause the position of the stars to apparently shift. Einstein even did the calculations to predict the direction they would shift and by how much. Unfortunately, the only mass in the solar system big enough to do such a thing was the sun. Trying to see the light of far off stars as it passes near the sun is akin to looking for fireflies near a large bonfire. The light is so intense that it obscures them. The only way to photograph the light of the stars was if the sun were somehow blocked out. What Einstein needed was a total solar eclipse.

Einstein appealed to the world's astrophysicists to help him prove his grand theory. They ignored him, all except for a young German astronomer named Erwin Finlay Freundlich. He was intrigued by Einstein's theory and excited to help find evidence. The two began corresponding, and after hurriedly consulting his astronomical charts, Freundlich determined that the next total solar eclipse would be visible in the Crimea, on August 21, 1914. Einstein was so thrilled that at first he offered to help pay for an expedition himself. Fortunately for his limited finances, they found a wealthy patron willing to pick up the tab. Nicely equipped, Freundlich and his associates set out with their telescopes and cameras for the Crimea on July 19, 1914.

There was only one small hitch, World War I. On August 1,

while Freundlich was still en route, Germany declared war on Russia. Shortly after they arrived in the Russian-controlled Crimea, the group of astronomers was discovered by the Russian army. The soldiers were rather disinclined to believe the band of Germans they discovered setting up camp with powerful telescopes and photographic equipment were there for innocent scientific purposes. They arrested the lot of them as spies, confiscated all their equipment and sent them off to a POW camp. Fortunately for Freundlich, the Russians released them and sent them home after a few weeks but refused to return the confiscated equipment.

The Great War enveloping Europe not only temporarily crushed Einstein's dreams of proving his beautiful theory, but also confirmed his deepest fears about human nature. He wrote to a friend, "Europe in its madness has now embarked on something incredibly preposterous, at such times one sees to what deplorable breed of brutes we belong." He watched in horror as this madness consumed even his closest friends.

Planck, Nernst, and a number of Einstein's fellow German scientists enthusiastically joined the war effort. Many of them became signatories to a proclamation titled Appeal to the Cultured World. It was a full-throated defense of German militarism and a denial that the Germans had committed any attacks on civilians. It came to be known as the "Manifesto of the 93" in honor of the number of prominent German intellectuals who signed it.

Einstein reacted to this nationalistic frenzy by working with physician Georg Friedrich Nicolai to draft a pacifist response. Known as the Manifesto to Europeans, it called on all civilized people to transcend petty nationalism and stated: "Such a mood

cannot be excused by any national passion; it is unworthy of all
that which the world has to date understood by the name of
culture. Should this mood achieve a certain universality among
the educated, this would be a disaster." Tellingly, it would go on
to predict:

> The struggle raging today will likely produce no victor; it will
> leave probably only the vanquished. Therefore, it seems not
> only good, but rather bitterly necessary that educated men of
> all nations marshal their influence such that—whatever the
> still uncertain end of the war may be—the terms of peace
> shall not become the wellspring of future wars.

It speaks volumes about the tenor of the time, that this pacifist appeal managed to have only four signatures, including Einstein's. Undeterred, Einstein would remain a committed and very
public voice for pacifism.

Ultimately, the cumulative effects of the war, the disintegra-
tion of his marriage, and his enormous effort to finish his general
theory of relativity became too much for Einstein. In 1917, he
suffered an almost complete physical collapse. He seemed to fade
to a shell of his former self. He lost a great deal of weight and
was so weak that he was unable to leave his apartment. His doctors diagnosed him as suffering from a severe stomach ulcer. Elsa
stepped in to take charge of nursing him back to health. She lovingly cooked a special diet and cared for his needs, eventually
moving him into the apartment next to hers.

By 1919, the war was over, and Einstein had recovered sufficiently to want to marry Elsa. Unfortunately, that required a formal divorce from Mileva, and she was hesitant. They had been

separated for five years, and she and the children were living in Zurich, while Einstein continued to live in Berlin, but Mileva was unwilling to make that final break.

After arguing back and forth over the matter for months, Einstein finally made Mileva an unbelievable offer. He was so confident in his own abilities, so certain of the revolutionary potential of his theories, that he told Mileva with absolute conviction that one day he would win a Nobel Prize. If she agreed to give him a divorce, he agreed to give her all of the winnings from the prize. At the time, the prize was worth 135,000 Swedish kronor or over forty-two thousand dollars in 1918 currency, approximately thirty-seven times the stipend she was getting from Einstein. Reluctantly, Mileva agreed. The following June, Einstein and Elsa were married.

Shortly before the wedding, on May 29, 1919, a new solar eclipse was to occur, just the chance Einstein needed to verify his theory. This time a British astronomer, Arthur Eddington led an expedition to the island of Principe, off the coast of West Africa, to capture the event. Eddington was particularly interested in relativity, and, as a Quaker and conscientious objector during the war, he believed it would be a grand gesture of postwar unity for a British scientist to confirm the theory of a German scientist. He nearly lost his chance, because on the appointed date the skies were overcast. As fate would have it though, right before the eclipse occurred, the clouds parted, and he was able to get the crucial photographs.

It took months to properly analyze the results, but on September 22, 1919, Einstein received a telegram informing him that the British team had pinpointed the stars deflected to almost exactly where he had predicted they should be. The difference was well within the limits of experimental error, and it was enough to confirm Einstein's theory of general relativity.

The scientific community was immediately set abuzz by the implications of Einstein's theory. The Royal Society and the Royal Astronomical Society held a joint meeting on November 6 to announce the results. The Astronomer Royal, Sir Frank Dyson presented the findings in a hall dominated by a large formal portrait of Isaac Newton. J. J. Thomson, the Nobel laureate and president of the Royal Society described it as, "one of the greatest achievements in the history of human thought. It is not the discovery of an outlying island but of a whole continent of new scientific ideas. It is the greatest discovery in connection with gravitation since Newton enunciated his principles."

The next day, the *Times* of London carried the headline: "REVOLUTION IN SCIENCE—New Theory of the Universe— NEWTONIAN IDEAS OVERTHROWN." When the news traveled across the Atlantic it was treated by the *New York Times* with similar drama: "LIGHTS ALL ASKEW IN THE HEAVENS— Men of Science More or Less Agog Over Results of Eclipse Observations—EINSTEIN'S THEORY TRIUMPHS."

Over the course of a few short days, Albert Einstein went from being an obscure, German professor to an international superstar. The war-weary public was captivated by the scientist and his ideas. The image of his wild hair and rumpled clothing, along with his eloquence and quick wit mesmerized them like no other scientist before or since. It was as if the public had fallen in love with the man, and even though they didn't understand his theory, they seemed to love the idea that his theories were so incomprehensible. As Einstein put it, "Now every coachman and waiter argues about whether or not relativity theory is correct."

Even other scientists had difficulty wrapping their heads around it. According to one story, when Eddington, the astronomer who confirmed Einstein's theory was leaving the Royal Society meeting where it was announced, he was stopped by a fellow scientist who told him, "There's a rumor that only three people in the entire world understand Einstein's theory. You must be one of them." When the astronomer paused but didn't say anything, the other scientist went on, "Don't be modest Eddington." Eddington simply looked at him and said, "Not at all. I was wondering who the third might be."

In 1921, Einstein came to America for the first time. He was there to talk about his theories and raise money for Hebrew University in Jerusalem. When his ship docked in New York, he was besieged by reporters. He and Elsa were treated to a motorcade in an open-topped limousine, and crowds lined the streets just to catch a glimpse of the famous scientist. Einstein asked his wife, "Do I have something of a charlatan or hypnotist about me that draws people like a circus clown?" Eight thousand people squeezed into the Sixty-Ninth Regiment Armory to hear him speak. Thousands had to be turned away.

When he went to Washington, D.C., Einstein met with President Warren G. Harding, while the U.S. Senate attempted to debate his theory of relativity. In city after city he was met by similar spectacles. Einstein endured the adulation with typical good humor. At a formal reception in the National Academy of Sciences, the famous scientist was obliged to sit through an interminable series of speeches and presentations. Eventually, he turned to the Dutch diplomat next to him and said, "I've just developed a new theory of eternity" (Isaacson, 2007).

Acceptance of Einstein's ideas, however, was not universal. A

Columbia University professor of celestial mechanics, Charles Lane Poor, attacked the theory and claimed that supposed astronomical proof did not exist. George Francis Gillette, a prominent engineer, publicly dismissed the entire idea as, "cross-eyed physics . . . utterly mad . . . the moronic brain child of mental colic . . . the nadir of pure drivel . . . and voodoo nonsense."

Even the editors of the *New York Times* became skeptical, proclaiming that the British people, "seem to have been seized with something like intellectual panic when they heard of photographic verification of the Einstein theory. . . . They are slowly recovering as they realize that the sun still rises—apparently—in the east." As Max Planck said when confronted with criticism of his quantum theory, "A new scientific truth does not as a rule prevail because its opponents declare themselves persuaded or convinced, but because the opponents gradually die out and the younger generation is made familiar with the truth from the start."

Criticism of Einstein was particularly acute in his native Germany. His fame made him a target for anti-Semitic forces. Notable among these was the Nobel Prize–winning physicist Philipp Lenard. Ironically it was Lenard's work that Einstein had used as the basis for his discoveries on light quanta. Now Lenard denounced Einstein as a "Jewish fraud." Lenard worked behind the scenes to deny Einstein a Nobel Prize, and became a leading spokesman for the Anti-Relativity League, an organization dedicated to establishing the purity of Aryan physics. By the beginning of 1921, this antirelativity rhetoric was being picked up by a Munich party functionary who wrote in a newspaper polemic, "Science, once our greatest pride, is today being taught by Hebrews." The functionary's name was Adolph Hitler.

By autumn 1922, the situation had become uncomfortable enough for Einstein that he decided to go on an extended tour of Asia. For the next six months he traveled the Orient, visiting Ceylon, Singapore, and Japan. On his voyage home, Einstein made a twelve-day stopover in Palestine. There he visited Tel Aviv, Haifa, and Jerusalem. While witnessing his fellow Jews industriously building a new land, Einstein declared, "Today, I have been made happy by the sight of the Jewish people learning to recognize themselves and to make themselves recognized as a force in the world."

It was while Einstein was on his Asian tour, that he was awarded the Nobel Prize. He had gotten strong hints that he would win before the trip, but the formal notification came in the form of a telegram sent to him on November 10 while en route to Japan, "Nobel Prize for physics awarded to you. More by letter."

It had been a long time coming. Einstein was first nominated in 1910, and oddly, when he finally won, it wasn't for his most famous work, the theory of relativity, but for his 1905 work on the photoelectric effect and light quanta. It must have been a blow for Lenard to realize that the Jewish physicist he had so long campaigned against, not only won the Nobel Prize but did so for work originally inspired by his own work on light. Einstein gave his acceptance speech when he returned from Asia and, as promised, gave the prize money to Mileva.

In 1905, Albert Einstein, along with Max Planck, had given life to one of the most revolutionary theories in the history of physics, quantum theory, later dubbed quantum mechanics. Even though he followed it up with the equally revolutionary theory of relativity, by the 1920s, Einstein began to wrestle with the implications of his earlier creation. Einstein's 1921 Nobel Prize was followed up with a prize for Niels Bohr in 1922, for his revised model of the atom. Bohr, along with a dedicated young band of colleagues, like Werner Heisenberg and Erwin Schrodinger, took up quantum theory where Einstein left off.

As Bohr and company expanded on the quantum theory, two things became apparent. First, it depended, at its most fundamental level, on random chance. Subatomic particles like electrons didn't behave like nice, logical bits of normal matter that adhered to the rules of classical physics. Rather, they behaved in strange unpredictable ways. Instead of knowing precisely where they are and what they're doing, theoretical physicists were left to make predictions about the probability that they were at a given location or had a given momentum, but one could never know for sure. Einstein found all of this deeply disturbing. If quantum mechanics were correct, and all the experiments seemed to suggest it was, then Einstein feared it might mean the end of physics and the concept of causality. This led to Einstein's oft repeated quote, "God does not play dice."

The second thing that became apparent was that quantum mechanics was incompatible with relativity. It was as if the universe had two sets of rules. At the level of people and planets and everything we can see, it played by the rules of relativity, but under the surface, at the subatomic level, there was an entirely new game operating with a different set of quantum rules. From the moment they first met, and for the remainder of their lives, Einstein and Bohr would carry on an intense but good-natured argument about this. In response to Einstein's quote about God not playing dice, Bohr responded, "Einstein, stop telling God what to do!" (Isaacson, 2007).

Hitler and the Nazis continued to ride the wave of postwar anger and anti-Semitic paranoia. By 1931, the situation in Berlin had deteriorated to the point that Einstein felt the need to leave. He came to America again for a two-month visiting professorship at the California Institute of Technology (Caltech). While there, he met American educator Abraham Flexner. Flexner, with the backing of Louis Bamberger and his sister Caroline Bamberger Fuld of the department store fortune, was in the process of setting up an extraordinary new institute. It was being established near Princeton University, although not affiliated formally with the school, and it was designed as a sort of haven where gifted scholars could work without the pressures of academic demands and teaching duties. It was called the Institute for Advanced Studies.

Once the visiting professorship was over, Einstein returned to Europe. While he was at Oxford, Flexner paid him a second visit and offered him a position at the institute. In the course of their discussions, Flexner asked Einstein how much he thought he should make. The scientist tentatively suggested three thousand per year. Flexner was quite amused, because he had much more in mind. The American told him, "Let Mrs. Einstein and me arrange it." By the time Flexner and Elsa were done, Einstein received an annual salary of fifteen thousand dollars. He was soon on his way to New Jersey.

Einstein arrived again in America, this time to stay, on October 17, 1933. Flexner immediately whisked him off to Princeton in an attempt to shield the famous physicist from the press. The attempt failed utterly. Einstein was soon granting interviews, attending benefits to raise money for his beloved pacifist and Zionist causes, and generally amusing the residents of the small New Jersey town with his antics. The rapidly expanding Einstein legend began including tales of the kindly but disheveled genius wandering around town in a baggy sweatshirt and no socks with his aurora of gray hair flying in all directions, completely lost in thought.

Sometimes this was more literal than others. One famous story recounts someone calling the institute. The unidentified caller asked to speak to a particular dean. When the secretary told him that he wasn't there, the caller asked if she could give him Dr. Einstein's home address. She of course said that she couldn't, at which point the caller whispered, "Please don't tell anybody, but I am Dr. Einstein. I'm on my way home, and I've forgotten where my house is" (Clark, 1971). The story may or may not be apocryphal, but like many others, it quickly spread.

In 1940, Albert Einstein became a U.S. citizen. He spent the remainder of his life in America, aside from occasional trips, the majority of it at Princeton. His time at the institute allowed him almost unlimited time to explore new ideas, but instead, he spent most of it trying in vain to poke holes in quantum mechanics. While a new generation of physicists was finding additional evidence for the radical theory, Einstein was increasingly cast in the role of stubborn traditionalist. He struggled in vain for some way to unify relativity and quantum mechanics. Perhaps he had seen this coming, because he had once remarked, "To punish me for my contempt of authority, Fate has made me an authority myself" (Isaacson, 2007).

Einstein was still questing after an elusive unified theory when he died at the age of seventy-six. In accordance with his wishes, the body was cremated a few hours after his death. A small private service was held for family and close friends, and then his ashes were scattered into the Delaware River. Einstein had no wish for his grave site to become a place of morbid pilgrimages and public mourning. Newspapers around the world noted the death and President Eisenhower declared, "No other man contributed so much to the vast expansion of 20th century knowledge. Yet no other man was more modest in the possession of the power that is knowledge, more sure that power without wisdom is deadly."

In a macabre development worthy of a mad scientist movie, one part of his body escaped cremation, Einstein's brain. The pathologist who performed the routine autopsy, Thomas Harvey, thought the organ would be of considerable value to science, and so, without asking permission, removed the brain and preserved it. When the family learned of this they objected strenuously, but Harvey managed to convince them that Einstein would have wanted his brain to be studied, so that others might learn. Reluctantly they acquiesced. Over the years, Harvey was besieged by researchers hoping to unlock the secret of Einstein's genius. The brain was carefully sliced and examined down to the cellular level, but the secret remained. Perhaps the key to unlocking it lay not in the brain itself, but in the words that it produced. In 1930 Einstein attempted to explain himself in an article titled, "What I Believe." In it he stated:

> The most beautiful emotion we can experience is the mysterious. It is the fundamental emotion that stands at the cradle of all true art and science. He to whom this emotion is a stranger, who can no longer wonder and stand rapt in awe,

is as good as dead, a snuffed-out candle. To sense that behind anything that can be experienced there is something that our minds cannot grasp, whose beauty and sublimity reaches us only indirectly: this is religiousness. In this sense, and in this sense only, I am a devoutly religious man.

CHAPTER SEVEN

Bombs and Rockets

Mad Science Goes to War

ON THE AFTERNOON OF OCTOBER 11, 1939, THE FINAN-
cier and economist Alexander Sachs stepped into the Oval Office
of President Franklin Delano Roosevelt. Sachs was acting as an
unlikely intermediary, bearing a letter of urgent importance from
Albert Einstein and his fellow physicist Leo Szilard. The scientists
wished to inform the president of the possibility of using uranium to create a nuclear chain reaction. Their letter explained the
enormous energy that would be released and the potential to use
such a device as a devastating new type of weapon. It contained
the ominous lines:

> This new phenomena would also lead to the construction of
> bombs, and it is conceivable—though much less certain—
> that extremely powerful bombs of a new type may thus be
> constructed. A single bomb of this type, carried by boat and

exploded in a port, might well destroy the whole port together with some of the surrounding territory.

As alarmed as the president was by this, he was further shaken by the concerns of both scientists that the Nazis were actively working on such a weapon. Roosevelt called in his personal assistant and declared, "This requires action" (Isaacson, 2007).

Thus the stage was set for the next chapter in the history of the mad scientist, and the history of the world itself. It wasn't the first time the creative brilliance of the scientific genius had been turned to the tools of war. Archimedes had set that precedent millennia before, but now it would be undertaken on a scale never before seen, and for stakes that couldn't be higher. The task of coordinating the effort fell to the most unlikely of scientists, a rail thin, elegant young theoretician plucked from the relative obscurity of academia and thrust into the mantle of the largest and most consequential project in history. In the end, he would succeed in harnessing the powers of the sun only to be consumed by the firestorm of postwar anticommunist hysteria. His name was J. Robert Oppenheimer.

J. Robert Oppenheimer was born in a fine stone house at 250 West Ninety-fourth Street in the Upper West Side of Manhattan on April 22, 1904. His father was Julius Oppenheimer, self-made Jewish German immigrant who made his fortune in New York's thriving garment trade. His mother, Ella Friedman, was a talented artist. Although both parents were Jewish, Robert, like Einstein, was raised in a secular household. Rather than a synagogue, his parents belonged to the Ethical Culture Society, an offshoot of American Judaism emphasizing the values of rational thought, humanism, and social justice. Its influence would mold young Robert's later life.

When Robert was seven, his parents enrolled him in the Ethical Culture School on Central Park West. The school had been a response to the demand for private schools open to the children of the rising Jewish middle and upper classes, and it embodied progressive core values, striving to instill in its students a belief in humanitarianism and social justice. Its motto was Deed, not Creed.

Robert thrived in its intellectual environment, and excelled in science as well as literature, ethics, and languages. He was soon reading the works of Homer and Virgil in their original Greek and Latin. One of Robert's teachers, Alberta Newton, described how, "He received every new idea as perfectly beautiful" (Sherwin). Robert was an exceptional student, able to skip several grades. For his part, Robert described himself during these years, "I was an unctuous, repulsively good little boy." He went on, "My life as a child did not prepare me for the fact that the world is full of cruel and bitter things."

Robert graduated in the spring of 1921, and spent the summer traveling through Germany with his parents. In the course of his travels, Robert came down with a serious case of dysentery and was so ill he had to be sent home on a stretcher. It took him nearly a year to recuperate, postponing his enrollment in Harvard.

By the following spring, Robert had recovered sufficiently for his father to decide it was time to send his son back into the world. One of Robert's former teachers, Herbert Smith, was planning to travel to the U.S. Southwest, and Julius thought the trip might be an excellent chance to toughen the boy and give him some valuable experience. He persuaded Smith to let Robert come along. It proved to be a life-changing time for the young man.

Robert learned to ride horses, and he and Smith traveled throughout New Mexico, spending days camping out and exploring the rugged mountain wilderness. Robert, the stick thin, once sickly Manhattanite, was soon impressing Smith with his endurance. One day, Robert and a few friends set out with packhorses for the Pajarito (Little Bird) Plateau. They rode up to a height of ten thousand feet and traveled to the Jemez Caldera, a twelve-mile-wide volcanic crater.

From there, they turned northeast and rode until they came to an isolated canyon with a stream running across its floor and cottonwoods growing alongside. These beautiful trees gave the canyon its Spanish name, Los Alamos. The only sign of human habitation was a remote boarding school. Robert fell in love with the rugged landscape. A few years later, while visiting from college, Robert remarked, "My two great loves are physics and New Mexico. It's a pity they can't be combined."

In the autumn of 1922, Robert entered Harvard, indulging in an eclectic selection of classes, ranging from philosophy to French literature to calculus. Eventually, he focused on becoming a chemistry major, and later switched to physics. His time in the West had allowed him to grow and fostered a newfound confidence in the young man, but the rarified atmosphere of Harvard reversed much of that. He became isolated and began to suffer bouts of depression.

Niels Bohr gave two lectures at Harvard in 1923, speaking on the rapidly changing views of the atom, and radiation. Robert attended both, and they had a profound effect on him. He was captivated by the famous physicist and his radical theories. In later years, Oppenheimer said, "it would be hard to exaggerate how much I venerate Bohr."

Robert graduated summa cum laude in 1925, completing his bachelor's degree in only three years, making the dean's list, and being selected for membership in the academic honor society Phi Beta Kappa. His future seemed secure when he was selected for a graduate position at Cavendish Laboratories in Great Britain. He would be studying with J. J. Thomson, winner of the 1906 Nobel Prize for discovery of the electron. With high hopes Robert boarded a ship for England.

Europe was a hotbed of physics in the 1920s. Bohr and his associates were exploring quantum mechanics and seemingly making new discoveries every day. Much of this cutting edge work had yet to make it across the Atlantic, and Robert found himself a bit behind the curve. As he struggled to catch up, he was relegated to the lowly job of making thin beryllium strips for use in the study of electrons. It was a painstaking, labor-intensive process, and while Robert would one day go on to be a brilliant theoretician, he had little talent in the laboratory. In a letter to one of his friends he confessed, "I am having a pretty bad time. The lab work is a terrible bore, and I am so bad at it that it is impossible to feel that I am learning anything."

Far from home and under great pressure, Robert again began suffering from depression. One day, his friend Jeffries Wyman visited Robert, and found him lying on the floor moaning and rolling from side to side. In another incident, Robert collapsed in the lab. His friends and family began to worry as Robert's behavior became increasingly erratic. His parents came to Cambridge to see if they could help, all to no avail. Eventually, this culminated in an episode that nearly got Robert kicked out of school, and came close to getting him arrested.

Robert allegedly tried to poison his tutor. Patrick Blackett, who would one day go on to win the Nobel Prize himself, was acting as the young American's tutor. Although Robert's motive remains unclear, he apparently poisoned an apple with chemicals from the laboratory and left it on Blackett's desk. Fortunately, Blackett never ate the apple, but the university authorities found out and threatened to expel Robert and file criminal charges.

Robert's parents, still visiting Cambridge, intervened with the university. Julius convinced the university that instead of expelling Robert, they should place him on probation on condition that he see a psychiatrist. Robert agreed and saw a succession of therapists. None of them seemed to have much success. As one psychiatrist familiar with the case described the problem, Robert "gave the psychiatrist in Cambridge an outrageous song and dance. . . . The trouble is, you've got to have a psychiatrist who is abler than the person who's being analyzed. They don't have anybody."

In spite of the efforts of both Robert and the psychiatrists, by the summer of 1926, he seemed to have recovered. He returned from a brief vacation in Corsica a changed man. His mood was brighter, and for whatever reason, he was able to relate better with others. He also began applying himself full-time to theoretical physics. This process was helped along when Robert met Bohr. He had been impressed when he heard Bohr speak, but meeting his idol in person catalyzed his desire to explore the world of quantum mechanics.

Robert had a chance that spring to visit the University of Leiden. He enjoyed his time with the German physicists and seemed to feel more at home with them than with his English classmates. He met Max Born, director of the Institute of Theoretical Physics at the University of Göttingen. Born, who had first coined the term *quantum mechanics,* was impressed with the young American and offered him a chance to study at Göttingen. Robert accepted. As he described it, "I felt completely relieved of the responsibility to go back into a laboratory. I hadn't been good; I hadn't done anybody any good, and I hadn't had any fun whatever; and here was something I felt just driven to try."

In Göttingen, Robert really came into his own. He thrived among his German colleagues and impressed his professors, immersing himself in quantum theory and soon making significant contributions to it. Born was so impressed that he coauthored a paper with him in 1927. "On the Quantum Theory of Molecules" is still considered a major breakthrough in the field and contributed to work in high-energy physics seventy years later. During this time, Robert was given the nickname Opje by his Dutch colleagues. When he came back to the States, it was anglicized to Oppie, and much to Oppenheimer's delight, the name stuck.

With his doctorate in hand, Oppenheimer returned to America, accepting a position at Caltech. They hired him specifically to introduce their students to the new physics coming out of Europe. Although he was brilliant and passionate about the subject matter, Oppenheimer was not the easiest of lecturers. His students often had a hard time keeping up. Eventually, he developed a unique style. Rather than meeting with students individually, Oppenheimer would have his graduate and postdoctoral students all meet at his office.

Each student had his or, in rare cases, her own desk, and Oppie would casually lean against the wall asking questions about each student's particular problem. He then solicited comments from the group. One of his former students, Robert Serber, described the sessions, "Oppenheimer was interested in everything, and one subject after another was introduced and coexisted with all the others. In an afternoon, we might discuss electrodynamics, cosmic rays and nuclear physics."

At the time, Oppenheimer seemed to be almost completely apolitical. He claimed not to have heard about the 1929 stock market crash until months after the fact and once asked a friend, "Tell me, what has politics to do with truth, goodness and beauty?" However, like many intellectuals, Oppenheimer had a number of friends on the political left, including his landlady, Mary Ellen Washburn, whose home acted as an informal social hub for Berkeley's literati, and Haakon Chevalier, a young professor of French literature. Years later, the FBI concluded that both had ties to the American Communist Party.

When Hitler came to power in January 1933, even Oppenheimer's practiced apoliticism fell by the wayside. Oppenheimer was asked to donate money to help Jewish physicists emigrate from Nazi Germany. Without hesitation, he made a sizable donation. As the situation in Germany deteriorated and the Depression dragged on, Berkeley, like many cities became increasingly polarized. Labor disputes and strikes were breaking out all over California, frequently ending violently. In the midst of this political maelstrom, the liberal values instilled in Oppenheimer's youth asserted themselves. He was soon attending rallies.

In July 1936, civil war broke out in Spain as the democratically elected leftist government came under attack by fascist forces of General Francisco Franco. The United States and the

European democracies initiated an arms embargo against both sides. Germany and Italy supported Franco's forces, and the Soviet Union lent aid to the besieged Spanish Republicans.

The plight of the Spanish Republicans quickly became a cause célèbre among American leftists. Oppenheimer, like many of his intellectual friends, actively supported it. He signed petitions, attended rallies, and made generous donations, many funneled through groups later labeled as communist front organizations by the House Un-American Activities Committee. Despite this, Oppenheimer never formally joined the Communist Party.

Oppenheimer, like many on the left, began distancing himself from the communists after the Soviet Union signed a nonaggression pact with Nazi Germany on August 24, 1939. Initially, Oppenheimer tried to downplay the importance of the pact as simply a Soviet means of self-preservation. However, his faith was further shaken by reports coming out of Russia about Stalin's brutal purges. His worst fears were confirmed when two of his physicist friends, George Placzek and Victor Weisskopf, who had actually been in the Soviet Union, visited Oppenheimer, and provided credible firsthand accounts of the terror taking place. Oppenheimer retained his faith in his core liberal values and the New Deal, and continued to associate with his friends on the left, but after that he cut many of his communist ties.

By early 1939, Oppenheimer, like many physicists, became aware of the possibility of creating a nuclear chain reaction. Shortly after Einstein and Szilard's letter was presented to Roosevelt in October, he learned that the president had authorized a Uranium Committee to explore the possibilities. Also in 1939, Oppenheimer's close friend at Berkeley, Ernest Orlando Lawrence, was awarded a Nobel Prize for his development of the cyclotron. As director of the university's radiation laboratory, Lawrence quickly began working on the nascent uranium project.

Lawrence was working secretly by 1941 on a process using his cyclotron to separate out uranium-235 isotopes, but he was running into difficulties and needed help. He wrote to his superiors and asked for Oppenheimer to be added to the project, telling them, "Oppenheimer has important new ideas." Anticipating security problems with his friend's radical political history, he added, "I have a great deal of confidence in Oppenheimer." His superiors agreed, and on October 21, 1941, Oppenheimer accompanied Lawrence to a meeting at the General Electric laboratory in Schenectady, New York. When a report on the meeting was sent to Washington, it included Oppenheimer's estimate that it would take 100 kilograms (about 220 pounds) of U-235 to produce an explosive chain reaction.

Around that time, the Roosevelt administration decided to kick production of an atomic bomb into high gear, replacing the Uranium Committee with a high powered group called the S-1 Committee which reported directly to the White House. The committee created a group to work on fast-neutron research at Berkeley. Lawrence continued to send reports emphasizing the importance of Oppenheimer's contributions, telling them, "He combines a penetrating insight of the theoretical aspects of the whole program with a solid common sense, which sometimes in certain directions seems to be lacking." Impressed by these reports, the S-1 Committee appointed Oppenheimer director of the research.

Oppenheimer began organizing a seminar of the finest minds in physics to begin working on a bare-bones design for an atomic weapon. Among those recruited were Hans Bethe, a German who had fled Europe in 1935; Felix Bloch and Emil Konopinski, both Swiss physicists working in the United States; and the Hungarian-born Edward Teller. He also invited a number of his former students, including Robert Serber.

Once the seminar began, Oppenheimer showed a deft hand at running the group. Teller later wrote of the experience, "As Chairman, Oppenheimer showed a refined, sure informal touch. I don't know how he acquired this facility for handling people. Those who knew him well were really surprised." Bethe was equally impressed, "His grasp of problems was immediate—he could often understand an entire problem after he had heard a single sentence. Incidentally, one of the difficulties that he had in dealing with people was that he expected them to have the same faculty."

In the fall of 1942, plans were made for a central weapons laboratory dedicated to producing the atomic bomb. Several members of the S-1 Committee proposed Oppenheimer to head the project. The army refused, citing security concerns. Their pick to head things up was Colonel Leslie R. Groves. He had a proven record as the point man for the Army Corps of Engineers' recently completed Pentagon. In preparation for the new project, the army promoted Groves to general. On September 18, 1942, he officially took charge of the Manhattan Engineer District, named for the location of its original offices. It quickly came to be known simply as the Manhattan Project.

Groves and Oppenheimer couldn't have been more different. Oppenheimer was tall and wispy thin, with a soft-spoken, eloquent manner. Groves was six feet tall, weighed 250 pounds, loud, plainspoken, and used to getting his way by muscling through resistance not by the niceties of debate or diplomacy. Oppenheimer was a theoretical scientist, used to seeing the big picture;

Groves was an engineer, used to diving in and getting his hands dirty. The scientist was a left-wing liberal who solicited the opinions of others; the general was a right-wing conservative, who expected his orders to be followed.

Despite their differences, from the first time they met, Oppenheimer and Groves seemed to have a grudging respect for each other. As Oppenheimer later summed up the general, "Oh yes, Groves is a bastard, but he's a straightforward one!" Groves was impressed by the scientist's intellect. As he once described Oppenheimer to a reporter, "He's a genius, a real genius. While Lawrence is very bright, he's not a genius, just a good hard worker. Why Oppenheimer knows about everything. He can talk to you about anything you bring up. Well, not exactly. I guess there are a few things he doesn't know about. He doesn't know anything about sports."

Each man thought he could work with the other. When Groves returned to Washington, he proposed to the Military Policy Committee that Oppenheimer be appointed director of the central laboratory. When they looked at Oppenheimer's file, including FBI surveillance reports on Oppenheimer and a number of his friends, the committee opposed his appointment. Groves responded, "After much discussion I asked each member to give me the name of a man who would be a better choice. In a few weeks it became clear that we were not going to find a better man." By the end of October, Oppenheimer had the job.

Groves was intrigued by Oppenheimer's idea of building the lab in an isolated location, rather than a major city. From Groves' point of view, the isolation made his job of maintaining tight security that much easier. Oppenheimer, along with fellow physicist Edwin McMillan and Major John H. Dudley began searching the Southwest for a suitable location in November 1942. They surveyed dozens of sites, without much success. Eventually, they investigated a canyon forty miles northwest of Santa Fe, but it proved to be too narrow for the size of the facilities they needed. Oppenheimer proposed the nearby site of Los Alamos, and the three men piled back into their car and drove the thirty miles across the Pajarito Plateau.

They eventually came to the boys school Oppenheimer had visited years before. The location featured eight hundred acres at an elevation of seventy-two hundred feet and included a main lodge building, dormitories, outbuildings, and a pond. It was exactly what they were looking for. When they brought Groves to see the site, he took one look and announced, "This is the place." Oppenheimer's dream of combining physics with his beloved New Mexico became a reality.

The army bought the school and began the Herculean task of building what amounted to an entire town from virtually nothing. The initial plan was to build a facility capable of housing the scientists, the support staff, and their families. It included bachelor quarters and homes for families, as well as a hospital, a library, a laundry, a school for young children, and an army exchange to act as grocery store and post office. There were also plans for a cantina, a mess hall, and a café where couples could eat out in the evening. The laboratory facilities included Van de Graaff generators, a cyclotron, and a Cockcroft-Walton machine, an early type of particle accelerator.

When the Los Alamos facility opened in March 1943, a hundred scientists, engineers, and support staff arrived. Within six months, that number had grown to a thousand. By the summer of 1945, the population had risen to four thousand civilians and two thousand men in uniform, housed in three hundred apartment buildings, fifty-two dormitories, and two hundred trailers. The Technical Area alone amounted to thirty-seven buildings, including a foundry, a plutonium purification plant, and an auditorium, plus dozens of labs, offices, and warehouses. Originally, the army budgeted $300,000 for construction. Within a year, they had spent $7.5 million.

Before Los Alamos, Oppenheimer had never supervised anything larger than a graduate seminar. Now he was responsible for one of the biggest scientific projects ever undertaken. He was at first ill-prepared and quickly ran into huge organizational problems. However, within a matter of months, he managed to transform himself from a somewhat eccentric leftist intellectual into an efficient administrator who instilled confidence in those under him. As Robert Wilson, one of the physicists who worked closely with Oppie put it, "He had style and he had class. He was a very clever man. And whatever we felt about his deficiencies, in a few months he had corrected those deficiencies, and obviously knew a lot more than we did about administrative procedures. Whatever our qualms were, why, they were soon allayed."

In addition to his administrative duties, Oppenheimer often had to run interference between the military and the scientists. Groves had initially proposed that all of the scientists be inducted into the Army as commissioned officers. Oppenheimer at first agreed, even subjected himself to an Army physical. He failed it. The reasons given were that he was seriously underweight, had back trouble and a chronic cough due to an old case of tuberculosis, aggravated by chain smoking. In spite of this, Groves pressured the doctors to pass him, and Oppenheimer was commissioned as a lieutenant colonel.

Oppenheimer's fellow scientists, however, balked at the idea of being put under military discipline. They insisted that they couldn't work under such conditions. After much wrangling, Oppenheimer managed to engineer a compromise. Groves agreed that during the experimental part of the job, the scientists would be considered civilians, but when it came time to actually test the bomb, they would put on uniforms. In addition, although Los Alamos was fenced and considered an army base, within the technical area, the scientists would report to Oppenheimer, and they would be free to exchange information with each other. When the negotiations were done, Hans Bethe congratulated Oppie, "I think that you have now earned a degree in High Diplomacy."

Once the bare necessities were in place, the hard work began. Scientists and technicians put in long hours, six days a week, pausing only on Sunday, and then back to work. Everyone was under the assumption that they were in a neck-and-neck race with the Germans, and the fate of the world rested in the balance.

The pressure only intensified when Oppenheimer learned that the Nobel Prize winner Werner Heisenberg was heading up the German atomic program. The army was so concerned that they considered plans to kidnap and/or assassinate Heisenberg when he visited Switzerland. Not wanting to tip their hand to the Nazis about how important the Allies considered the atomic program, the plan was scrapped.

Despite the pressure, the lack of amenities, and the enforced isolation, Oppenheimer continually inspired his people to do their best. Victor (Vicki) Weisskopf, one of the scientists summed it up, "He was present in the laboratory or in the seminar room when a new effect was measured, when a new idea was conceived.

It was not that he contributed so many ideas or suggestions; he did so sometimes, but his main influence came from his continuous and intense presence, which produced a sense of direct participation in all of us." This was a common feeling among those at Los Alamos. Wilson, who initially had some reservations about his new boss, put it this way, "when I was with him, I was a larger person. . . . I became very much of an Oppenheimer person and just idolized him. . . . I changed around completely."

On December 30, 1943, Oppenheimer had an important visitor. Niels Bohr came to Los Alamos. He had been smuggled out of Copenhagen a few months before and transported secretly to England. Heisenberg had been one of Bohr's students, and in late 1941 the Danish scientist had met with his German protégé in Copenhagen. Now Bohr was able to confirm for Oppenheimer that the Germans were working on an atomic bomb, but he was also able to assure the Americans that the Nazis were not as far ahead as they feared.

This gave Oppie and the others a great deal of encouragement, but it hadn't been Bohr's primary reason in coming. What the older scientist wanted to do was get Oppenheimer thinking about the implications of the atomic bomb after the war. Bohr was greatly concerned, and rightly so, about the possibility of a postwar atomic arms race between the United States and the Soviet Union. He wanted Oppenheimer and the others to consider the ethics of what they were doing. Bohr later said, "That is why I went to America. They didn't need my help in making the atom bomb."

Bohr left Los Alamos in the spring, and the work continued. Everyone knew they had monumental problems to solve and not much time to do it. There were constant struggles to come up with sufficient amounts of fissionable material, and the scientists were kept busy trying to improve methods of producing it. There was also the question of whether the bomb should use U-235 or plutonium. Both designs were explored. They also debated the best method of triggering the chain reaction. One group argued for a gun-type design that would fire a slug of fissionable material into the core to start the reaction. Another group proposed an implosion design where plutonium was surrounded by conventional explosives. Detonating those would drive the plutonium atoms together and start the chain reaction. Oppenheimer was forced to mediate between all the competing groups.

Everything changed on April 12, 1945, when Franklin Delano Roosevelt died. Work was temporarily suspended, and Oppenheimer assembled everyone to make a formal announcement. That Sunday at Los Alamos, a memorial service was held. Oppenheimer delivered the eulogy and began, "We have been living through years of great evil, and great terror." He went on to describe Roosevelt as, "in an old and unperverted sense, our leader." The scientists concluded his tribute with, "we should dedicate ourselves to the hope, that his good works will not have ended with his death." Later, Oppenheimer confided in one of his friends, "Roosevelt was a great architect, perhaps Truman will be a good carpenter."

Later that month, on April 30, Adolph Hitler committed suicide. Germany formally surrendered eight days later. While the war in the Pacific raged on, the war in Europe was over. For two years, the people at Los Alamos had worked and sacrificed under the unwavering belief that they were building this terrible weapon to keep the Nazis from winning. No one, not even the military, considered the possibility that Japan had an atomic program. As the scientists began to privately discuss what to do next, Bohr's call to think about the implications of what they were doing was renewed.

A group of atomic scientists working in Chicago organized an informal committee and drafted a twelve-page report on the social and political implications of the bomb. It came to be known as the Franck Report, after the group's chairman, Nobel Prize winner James Franck. In no uncertain terms it argued against using the atomic bomb for a surprise attack on the Japanese: "It may be very difficult to persuade the world that a nation which was capable of secretly preparing and suddenly releasing a weapon as indiscriminate as the [German] rocket bomb and a million times more destructive, is to be trusted in its proclaimed desire of having such weapons abolished by international agreement." The report concluded that if the atomic bomb were to be used, it should be demonstrated for representatives of the United Nations, perhaps in the desert or an isolated island. They forwarded the report to the president. It was seized by the army and immediately classified. Truman never saw it.

The discussions at Los Alamos were equally heated. In mid-June, Oppenheimer called for a meeting of the top scientists to give their recommendations. Everyone was aware of the Franck Report and what followed was an open and freewheeling discussion. When it was done, Oppenheimer prepared a summary for Secretary of War Henry Stimson. First, it included the group's conclusion that before any use of an atomic weapon, the Allies, Britain, France, Russia, and China, should be informed. Second, the report concluded that while there was no unanimity, many of those present favored the idea of a demonstration.

Oppenheimer had argued for the military use of the bomb as a means of ending the war quickly and saving American lives. He explained that he and some of his colleagues believed use of the bomb might actually, "improve the international prospects, in that they are more concerned with the prevention of war than with the elimination of this specific weapon. We find ourselves closer to these latter views; we can propose no technical demonstration likely to bring an end to the war; we see no acceptable alternative to direct military use."

Two weeks after Oppenheimer sent the summary to Stimson, Teller showed Oppie a copy of a petition being circulated by Szilard, whose original letter with Einstein had started the whole thing. Szilard's petition urged Truman not to use the atomic weapon on Japan without making public the terms necessary for their surrender and giving the Japanese a chance to surrender. The petition had been signed by 155 Manhattan Project scientists. A separate poll of the project scientists conducted by the army, found 72 percent favored a demonstration rather than its military use without prior warning.

According to Teller, when Oppie saw the petition, he was angered and said, "What do they know about Japanese psychology? How can they judge the way to end a war?" Oppenheimer, however, was unaware of much that was happening behind the scenes. Truman had gotten a promise from Stalin that Russia would declare war on Japan on August 15. Truman wrote in his diary on July 17, "Fini Japs when that comes about."

In addition, the military had intercepted messages from Japan indicating that the Tokyo government was looking for ways to end the war. Truman's chief of staff, Admiral William D. Leahy, wrote in his diary on June 18, "It is my opinion at the present time that a surrender of Japan can be arranged with terms that can be accepted by Japan."

When General Dwight D. Eisenhower, was informed in July of the existence of the atomic bomb, he told Secretary of War Stimson that he thought it wasn't needed, because "the Japanese were ready to surrender and it wasn't necessary to hit them with that awful thing."

In spite of this, the work at Los Alamos continued. Groves put additional pressure on everyone to have the bomb completed before August 10. He didn't say why, but insisted that was the deadline they "had to meet at whatever cost in risk or money or good development policy." On July 16, 1945, they were ready to test "the gadget" as it had come to be known.

Oppenheimer sent his brother, Frank, and Ken Bainbridge to make final preparations at the test site dubbed Trinity. For several weeks, they slept in tents and worked in hundred-degree-plus temperatures to make sure that everything was ready. The name "Trinity," for the site near Alamogordo, had been picked by Oppie. It was inspired by a poem by John Donne. The night before the test, July 15, the weather looked bad. Neither Groves nor Oppenheimer wanted a delay, so it was decided that the test would commence at 5:30 A.M., and they would just have to hope for the best.

At 5:00 A.M. the skies started to clear, and at 5:10 the voice of physicist Sam Allison came over the loudspeaker to announce, "It is now zero minus twenty minutes." As the final seconds ticked down, everyone scrambled into position and grabbed protective eye gear. Joe Hirschfelder, one of the chemists assigned to measure the radioactive fallout described what happened:

All of a sudden, the night turned into day and it was tremendously bright, the chill turned into warmth; the fireball gradually turned from white to yellow to red as it grew in size and climbed into the sky; after about five seconds the darkness returned but with the sky and the air filled with a purple glow, just as though we were surrounded by an aurora borealis. . . . We stood there in awe as the blast wave picked up chunks of dirt from the desert soil and soon passed us by."

Oppenheimer, who had been lying on the ground only ten thousand yards from ground zero, looked up when it was done and said simply, "It worked." In a famous television interview to NBC in 1965, Oppie described the scene in more poetic terms:

We knew the world would not be the same. A few people laughed, a few people cried. Most people were silent. I remembered the line from the Hindu scripture, the Bhagavad-Gita; Vishnu is trying to persuade the prince that he should do his duty, and to impress him, takes on his multi-armed form and says, "Now I am become death, the destroyer of worlds." I suppose we all thought that one way or another.

After the test, the mood back at Los Alamos was jubilant. People had worked hard, had dedicated their lives to this accomplishment. All the tension building for two years suddenly cut loose. Now it was time to celebrate. Scientists, technicians, soldiers—everyone joined in, drinking, dancing, and slapping each other on the back. The next day, people returned to work, but the feeling of elation continued. At least it continued for everyone but Oppenheimer.

Over the next several days, he was busy in meetings with U.S. Army Air Force officers trying to pick targets. Oppie knew each of the Japanese cities on the list and that knowledge sobered him. A few days later, Wilson saw him walking to work at the Technical Area, as usual puffing on his pipe. As Wilson walked up to him, he heard Oppenheimer muttering, "Those poor little people, those poor little people."

On the morning of August 6, 1945, the B-29 aircraft, nicknamed the Enola Gay, after the pilot's mother, approached the city of Hiroshima. At precisely 8:14 A.M. they dropped the bomb, code named "Little Boy." Due to communication delays, it took five hours for the news to reach Washington, D.C., Captain Parsons, the arming officer on the Enola Gay sent a Teletype describing the attack, "The visible effects were greater than the New Mexico test."

The initial death toll in Hiroshima was estimated at seventy thousand (U.S. Department of Energy). Those are the people who died instantly when the bomb detonated or shortly thereafter. Others died over the course of several days or weeks from burns and radiation poisoning, pushing the death toll above one hundred thousand. Long-term damage from radiation, such as cancer, effectively doubled that figure over the next five years.

Two days later, the Soviets officially declared war on Japan and invaded Manchuria. The next day, August 9, a second atomic bomb, code named "Fat Man" was dropped on the city of Nagasaki at 11:02 A.M. Parts of the city had been evacuated due to previous air raids. As a result the death toll was somewhat lower than Hiroshima, but the psychological effect was just as devastating. The Japanese surrendered unconditionally on August 14, 1945.

If Oppenheimer had doubts about his achievement, the public didn't. Oppenheimer became a national hero. His face was soon splashed on newspapers and magazines across the country. *Scientific Monthly* declared, "Modern Prometheans have raided Mount Olympus again and have brought back the very thunderbolts of Zeus." *Life* magazine proclaimed that physicists now seemed to wear "the tunic of Superman." The *St. Louis Post-Dispatch* ran an editorial stating that never again should America's, "scientific explorers . . . be denied anything needful for their adventures." In the minds of the American public, Oppenheimer was the father of the atomic bomb.

Oppie used his newfound fame to call for international controls on nuclear weapons. When he resigned his position at Los Alamos in October, he gave a farewell address to his colleagues and their families. In the speech, he expressed his hope that in the future they would all look back on what they had achieved with pride, but he also warned, "Today that pride must be tempered with a profound concern. If atomic bombs are to be added as new weapons to the arsenals of a warring world, or to the arsenals of nations preparing for war, then the time will come when mankind will curse the names of Los Alamos and Hiroshima."

Oppenheimer knew when he left Los Alamos that his notoriety would grant him access to the doors of power in Washington. He used that access to campaign against the type of nuclear arms race he and Bohr feared. As part of that campaign, Oppenheimer was one of the driving forces behind the establishment of the Atomic Energy Commission (AEC) on August 1, 1946. He became one of its advisers, and at first, he was optimistic that a civilian agency like this could help control the atomic monster that he helped unleash. That optimism turned to disappointment in the face of American paranoia and Soviet intransigence.

Oppenheimer continued to speak out. He warned of the dangers of atomic weapons in newspapers, magazines, even the newly emerging television. His outspokenness earned him many admirers, but it also made him many enemies, some of them quite powerful. At Los Alamos, Edward Teller had promoted the idea of what he called a super bomb. Today, we call it the H-bomb, and it offered to create destruction orders of magnitude greater than the A-bomb. Oppenheimer had dismissed it as impractical and too time-consuming.

After the war, Teller continued to advocate his new weapon. Once the Soviets gained their own atomic bomb, Teller, a fanatical anticommunist thought that the super bomb was the only way to guarantee American superiority. Oppenheimer, and many others, saw the weapon as being an instrument of genocide with no legitimate military use. The AEC listened to Oppenheimer and recommended that the H-bomb not be pursued. Truman overruled them and went ahead with the program anyway, but Teller never forgot Oppie's opposition to what Teller saw as his bomb.

Meanwhile, Oppenheimer decided to go back East. Lewis Strauss, one of the trustees of the Institute of Advanced Studies in Princeton, offered Oppie the position of director of the institute. After giving the matter some thought, Oppenheimer accepted. In 1933, the institute had become the home to Albert Einstein, the world's most famous physicist. Now Oppenheimer was, at least nominally, becoming Einstein's boss. Unfortunately, Oppie's role at the institute was primarily an administrative post, and allowed him to do little actual physics. Nevertheless, Oppenheimer deftly handled his new duties and maintained the institute's reputation for attracting the best minds in the world.

At first, Oppenheimer and Strauss seemed to get along well enough, but it didn't take long for tension between them to grow. Initially, there were small clashes about institute matters, like appointments and budgetary matters, but these soon escalated. Both men were trying to assert their dominance. Although a businessman, Strauss was greatly interested in atomic power. When the AEC was formed, Truman made Strauss one of the first five commissioners. Unlike Oppenheimer, Strauss supported the creation of a hydrogen bomb. When Oppenheimer convinced the majority of the AEC to recommend against the H-bomb, Strauss, like Teller, took it as a personal affront.

The final straw for Strauss occurred when Oppenheimer testified on Capitol Hill in June 1949. The Joint Committee on Atomic Energy was holding an open session on the export of radioisotopes. Strauss had vehemently opposed it as a serious breach of national security. Oppenheimer supported the exports and, in his testimony, made some humorous off-the-cuff remarks at Strauss' expense. The audience and the committee members laughed. Strauss did not. David Lilienthal described the look on Strauss' face: "There was a look of hatred there that you don't see very often in a man's face."

Strauss was one of the most powerful men in Washington. An arch-conservative with many important contacts, he did not take ridicule lightly. One of his fellow AEC commissioners said, "If you disagree with Lewis about anything, he assumes you're just a fool at first. But if you go on disagreeing with him, he concludes you must be a traitor." He knew of, and was appalled by, Oppenheimer's leftist background, and now, in the midst of the anti-communist hysteria that Senator Joseph McCarthy had whipped up, Strauss had a chance to use it against the scientist.

Strauss went to J. Edgar Hoover. The FBI had been keeping tabs on Oppenheimer since the 1930s. When Oppie was appointed to Los Alamos, the FBI kept him under total surveillance, tapping his phone, bugging his house. Even his driver and body guard was an agent. Hoover shared with Strauss everything he knew about Oppenheimer, including his contributions to what were deemed communist front organizations and his close association with known communists. Neither Strauss nor Hoover was able to prove that Oppenheimer had done anything criminal, but as adviser to the AEC, Oppie needed a security clearance. That was the wedge Strauss used against him.

In 1954, Oppenheimer was presented with a series of charges and told his security clearance was going to be revoked. Oppenheimer demanded a hearing, playing directly into Strauss' hand. Strauss was chairman of the AEC and had the power to set the conditions for the hearing, including appointing the judges. If that weren't enough of a stacked deck, Strauss requested that Hoover bug not only Oppenheimer's house but the offices of his lawyers as well. That meant that Strauss' hand-picked prosecutor knew what the defense strategy was going to be. The defense, on the other hand, was hobbled, when Strauss managed to have Oppie's defense lawyers denied security clearances. They couldn't even look at most of the material being used against their client.

The hearing of the AEC Security Board took place in April 1954. Roger Robb, an experienced trial lawyer and assistant U.S. attorney acted as prosecutor. Over the course of the next three and a half weeks, he grilled Oppenheimer. In total, the scientist spent twenty-seven hours in the witness chair. When the scientist was done, the prosecution called witnesses to testify against him. One witness was Teller, who testified that Oppenheimer had deliberately endangered national security by obstructing the building of the hydrogen bomb.

Oppenheimer's lawyers were never able to mount an effective defense. In the end, the verdict was a foregone conclusion. They voted two to one to revoke Oppenheimer's security clearance. Ironically, after all the courtroom drama and behind-the-scenes machinations, the verdict came down the day before Oppenheimer's security clearance was due to expire. The entire affair may have been best summed up by Einstein, who said, "The trouble with Oppenheimer is that he loves a woman who doesn't love him—the United States government."

Oppenheimer returned to Princeton, and even though he continued as head of the institute, he was never the same after that. The ordeal of the trial and having his patriotism questioned seemed to take a lot out of him. Years later, on December 2, 1963, Oppenheimer was awarded the Enrico Fermi Prize by President Lyndon Banes Johnson. It was awarded for "excellence in research in energy science and technology benefiting mankind." There could be no mistake that it was a highly symbolic rehabilitation for the scientist who had done so much for his country. Hans Bethe summed up his friend's contribution this way:

> Los Alamos might have succeeded without him, but certainly only with much greater strain, less enthusiasm, and less speed. As it was, it was an unforgettable experience for all the members of the laboratory. There were other wartime laboratories of high achievement . . . But I have never ob-

served in any one of these other groups quite the spirit of belonging together, quite the urge to reminisce about the days of the laboratory, quite the feeling that this was really the great time of their lives. That this was true of Los Alamos was mainly due to Oppenheimer. He was a leader.

Even while the Americans were working feverishly to create the atomic bomb, their counterparts in Germany were working just as hard to weaponize the fruits of science. The Nazis had their own genius. He was young and charismatic, tall, blond, and blue-eyed, everything the Führer could want. He was also a man of vision, but that vision wasn't of world conquest. It was of conquering the frontier of space. Toward that end, he dedicated his life, and what a life it was, full of narrow escapes and devastating explosions, political intrigue, fame, power, and Faustian bargains. His name was Wernher von Braun.

Wernher Magnus Maximilian Freiherr von Braun was born on March 23, 1912. He was the second son of the Baron Magnus von Braun and the Baroness Emmy von Braun. Upon birth, Wernher became a member of the landed aristocracy, from a noble Prussian family. His early years were spent on the family's estate in Wirsitz, then part of Prussia, now Poland. When Wernher was still a toddler, however, his life was disrupted by the outbreak of The Great War, what would come to be known as World War I.

Magnus von Braun served the Kaiser faithfully during the war and, even after Germany's defeat, was able to maintain much of the family's wealth and influence. He soon got a job in the new Weimar Republic government, and the entire family moved to Berlin. It was there that Wernher and his two brothers, Sigismund the oldest, and Magnus the youngest received the finest education available at the French Gymnasium. There the boys learned French, Latin, Greek, and English. Unfortunately, like most gymnasiums, science was not emphasized. Wernher would have to learn that on his own.

While all the von Braun boys were very bright and excelled at school, Wernher showed that he had a special gift. His mother, who was very accomplished herself, described her son, "He was like a dry sponge, soaking up every bit of knowledge as eagerly as he could. There was no end of the questions he asked; he really could wear you out quickly! I never succeeded in being cross with him. . . . Whenever I tried to, he would put on his most cherubic smile and talk about something else" (Neufeld). His cherubic smile would come in handy later for getting him out of trouble.

When Wernher was twelve, he and Sigismund tried to learn about science by conducting their own experiments. The two boys were fascinated by rocket-powered automobiles that they read about in magazines, and they decided to build one themselves. They bought six of the largest skyrockets they could find and attached them securely to Wernher's wagon. They then wheeled the vehicle out onto Tiergarten Strasse, a large nearby street. They lit the fuses, and Wernher hopped in. In later years, Wernher described what happened next:

> I was ecstatic. The wagon was wholly out of control and trailing a comet's tail of fire, but my rockets were performing beyond my wildest dreams. Finally they burned themselves out with a magnificent thunderclap and the vehicle rolled to a halt. The police took me into custody very quickly. Fortunately, no

one had been injured, so I was released in charge of the Minister of Agriculture—who was my father (Braun, 1958).

As his brother pointed out, Wernher's account conveniently neglected to mention the fate of the fruit stand he collided with. Their father paid for the damages and after delivering a stiff lecture, grounded the two budding experimentalists for a couple of days. The boys' mother was somewhat more sympathetic. She told her two sons, "The world needs live scientists, not dead ones."

Shortly after this, Wernher was sent to a prestigious boarding school by his parents. While Wernher was away, his mother sent him a gift that would have a profound effect on his life. She sent him a small telescope. He was thrilled, and spent many hours using it to gaze at the heavens. Wernher described the effect this had on him, "I became an amateur astronomer, which led to my interest in the universe, which led to my curiosity about the vehicle which will one day carry a man to the Moon."

Around the age of fifteen, Wernher began reading science fiction. The adventurous stories and colorful illustrations fed his enthusiasm. After reading one in particular about going to the Moon, Wernher said, "It filled me with a romantic urge. Interplanetary travel! Here was a task worth dedicating one's life to! Not just to stare through a telescope at the Moon and the planets, but to soar through the heavens and actually explore the mysterious universe! I knew how Columbus had felt."

If the science fiction stories excited him, imagine Wernher's reaction when he saw a nonfiction book about space flight. He found a mention of the book *Die Rakete zu den Planeten-*

raumen (The Rocket into Interplanetary Space), by the German scientist Hermann Oberth, in one of his science fiction magazines. The teenager immediately ordered a copy. Day after day, Wernher anxiously checked the mail. He was so filled with excitement when the book finally arrived, that he ran with it up to his room, ripped off the wrappings and opened the book to learn its mysteries.

To his horror, what he found was page after page of complex mathematical formulas that left him completely baffled. He went to his teachers and desperately asked them how he could understand this book. They told him to study mathematics and physics. Up until that point, they had been his worst subjects, but with the goal of learning how to fly rockets to the moon, he now undertook them with a vengeance.

Wernher graduated with honors in 1929 at the age of seventeen. He returned to Berlin, and was thrilled when he learned that his idol, Professor Oberth, was in the German capital as well. Originally from Transylvania, part of Romania, Oberth had written about the possibility of going into space in rocket powered ships and was responsible for making space travel a popular topic in Germany at the time. He came to Berlin to act as a technical adviser for the hit film *Frau im Mond* (Woman in the Moon), and had managed to attract a band of young rocket enthusiasts. They formed a group called the Verein für Raumschiffahrt (Society for Spaceship Travel), or VfR. Wernher von Braun soon joined them.

In the spring of 1930, von Braun enrolled in the Charlottenburg Institute of Technology to study mechanical and aircraft engineering. When he wasn't in classes, von Braun was acting informally as Oberth's apprentice. More often than not, this consisted of raising money to continue their rocket experiments. Oberth and von Braun would set up a little display in one of the local department stores, and von Braun would solicit donations from passing shoppers, telling them, "I bet you that the first man to walk on the Moon is alive today somewhere on this Earth!" Unbeknownst to von Braun, Neil Armstrong had been born a short time before in Ohio.

Of course, there was more to assisting Oberth than begging for donations. Frequently von Braun and the others would get to help the professor with his experiments. This was both thrilling and not a little dangerous. Von Braun described how he and one of his fellow enthusiasts, Klaus Riedel, helped test the rocket engines, "Riedel lighted a rag soaked in gasoline. We opened valves that let the propellants into our tiny motor. Riedel hurled his torch over the motor's mouth and ducked behind a barricade. The jet ignited with a thunderclap that turned into a roar. After 90 seconds our fuel was exhausted and so were we."

Oberth returned to Transylvania later in 1930, but von Braun and the VfR continued their experiments, becoming very adept at scrounging for money and materials. In the midst of the Depression, they could occasionally recruit unemployed mechanics and technicians to help in exchange for free meals. They managed to find space to conduct their work by convincing the local authorities to let them use an abandoned ammunition storage depot on the outskirts of town for a nominal rent. They named the site Raketenflugplatz Berlin (Rocket Flight Field Berlin).

Customarily, European university students spent a semester or two at another institution to broaden their education. Von Braun did so in the spring of 1931, going to the Federal Technical University of Zurich. While there, he met a young American medical student named Constantine D. J. Generales. The two became friends, and while telling Generales about his love of rocketry, von Braun showed his new friend a written response to a letter he had written to Albert Einstein. The response was favorable and full of formulas that might be useful for the budding rocketeer.

Soon, von Braun had recruited Generales to help. The medical student suggested that before going into space it might be helpful to find out what the effects of space travel would be on the body. To do so, they got some white mice, and von Braun rigged up a home-made centrifuge out of bicycle parts. They would put the mice in tiny slings attached to the bicycle wheel, and when they turned the handle it would spin the mice around to simulate the G-forces experienced when a rocket blasted off.

The results were less than encouraging. A number of the mice died from the effects of the acceleration. Worse, occasionally a mouse would come loose from its sling and get flung across the room. When the landlady started discovering blood splattered spots on her walls, she became enraged and threatened to call the authorities and kick the two of them out. The students wisely decided to suspend their experiments for the time being.

Von Braun returned to Berlin. His friends in the VfR had continued their experiments and were starting to enjoy some success. Over the remainder of that year and the next, he and the group managed eighty-five rocket launches. Some reached altitudes of over twelve hundred feet. Word of their successes spread, and soon they were hosting visitors who would occasionally even pay, to see the blastoffs. In 1932, however, they had a different sort of visitor. Von Braun described the incident years later to *New Yorker* writer Daniel Lang:

One day in the spring of 1932, a black sedan drew up at the edge of the Raketenflugplatz and three passengers got out to watch a rocket launching. "They were in mufti [civilian clothing] but mufti or not, it was the Army," von Braun said to me. "That was the beginning. The Versailles Treaty [post-World War I] hadn't placed any restriction on rockets, and the Army was desperate to get back on its feet. We didn't care much about that, one way or the other, but we needed money, and the Army seemed willing to help us. In 1932, the idea of war seemed to us an absurdity. The Nazis weren't yet in power. We felt no moral scruples about the possible future abuse of our brainchild. We were interested solely in exploring outer space. It was simply a question with us of how the golden cow could be milked most successfully" (Lang, 1951).

One of the visitors was Army Captain Walter Dornberger, a career military man with a degree in mechanical engineering. He was impressed by the young group, and awarded them a thousand-mark contract to get better equipment. Happily, they took the money, and went to work. By July, they were ready for a demonstration at the Kummersdorf proving grounds. Before an audience of army officers, they launched the rocket. Majestically it rose about two hundred feet into the air. It then began to veer wildly and crashed spectacularly into the ground. The disappointment of the officers was evident.

Von Braun, however, would not let a simple crash hold him back. He gathered all of the data from the sensors and took them to Colonel Karl Becker, army chief of ballistics and ammunition. Von Braun passionately made his case, explaining that all they needed was some additional funding so they could become more professional in their work. The colonel seemed more impressed with von Braun than the rocket demonstration. He agreed, but made the funding conditional on the rocketeers working in secret within the confines of an army installation. Von Braun took the news back to his friends. Not all of them were thrilled by the conditions, but they recognized the need for funding, and they agreed.

The rocket society was under contract to conduct rocket R&D for the army. By the autumn of 1932, von Braun was appointed the group's civilian head at the tender age of twenty. At the same time that he was acting as chief of a classified research program, von Braun, with the help of Captain Dornberger, enrolled in the Friedrich-Wilhelms University in Berlin to pursue his doctorate degree. His doctoral research was an extension of his rocket research, and when he delivered his thesis in 1934, it was given the innocuous sounding name "About Combustion Tests," and immediately classified. At the age of twenty-two, he became Dr. Wernher von Braun.

Only a year earlier, in January 1933, Adolph Hitler came to power. Winning a simple plurality of the vote, he was appointed chancellor. Not long after that, the Baron Magnus von Braun took the opportunity to retire from government service. The younger von Braun seems to have been untroubled by the developments. According to von Braun's fellow rocket enthusiast Willy Ley, "Did we discuss politics? Hardly, our minds were always far out in space." Whether that statement is completely true has been debated by historians, but there's no evidence that at the time von Braun took a strong political stand one way or the other.

Just before Christmas 1934, their experiments seemed to pay

off. They successfully launched a pair of liquid-propelled rockets to a height of over a mile and a half. Officially, the rockets were designated A-2 (A for aggregate), but informally von Braun and his crew nicknamed them "Max" and "Moritz," after the German cartoon characters who would inspire the American Katzenjammer Kids.

The success of the A-2 launches soon attracted the attention of the German Air Force, the Luftwaffe. They offered the von Braun team five million reichsmarks (approximately $1.2 million) to come and work for them on a rocket-powered fighter plane. Alarmed at the possible loss of their valuable rocket researchers, the army countered by offering six million reichsmarks. The bidding war had begun. In the words of von Braun, rocketry had "emerged into what the Americans call the 'big time'" (Ward, 2005). They ended up staying with the army, but it wouldn't be the end of the struggle to control von Braun and his precious rockets.

Von Braun, like all able-bodied German men, was required to serve two years of military service. Ironically, even though he was working for the army, he did his service in the Air Force, becoming an accomplished pilot and eventually earning his qualification to fly military aircraft. He retained his love of flying for the rest of his life and indulged it whenever possible. Perhaps von Braun saw his pilot's training as bringing him that much closer to space.

By 1935, pressure was being exerted on the program to produce usable weapons. Dornberger, since promoted to major, told von Braun, "The Ordnance Department expects us to make a field weapon capable of carrying a large warhead over a range much beyond that of artillery. We can't hope to stay in business if we keep on firing only experimental rockets." Their group already numbered about twenty, and was rapidly running out of room, so Dornberger and von Braun began looking for a larger facility.

Von Braun mentioned the situation while visiting his parents over Christmas. His mother helpfully suggested, "Why don't you look at Peenemunde. Your grandfather used to go duck-shooting there." Von Braun was not only a brilliant rocket scientist; he was smart enough to listen to his mother. Peenemunde, on the island of Usedom, just off the Baltic coast, was exactly what they were looking for.

What followed was a bizarre parallel to Los Alamos. Over the course of two years, the sandy woods of Peenemunde were converted into a working town with housing, laboratories, factories, power plants, fire stations, and stores. It would eventually be home to thousands of scientists, technicians, and workers. Unlike Los Alamos, by the end of the war, it would also contain a forced labor camp for several thousand POWs and other prisoners. Effectively, the base was split in two. Peenemunde West was controlled by the Luftwaffe, and Peenemunde East was controlled by the army. Von Braun was the technical director of the army side. Dornberger, now a colonel, was the army commander.

They moved into their new accommodations at Peenemunde the month after von Braun's twenty-fifth birthday. The blond-haired, baby-faced von Braun was in charge of the research program working on the A-4 rocket. It was a forty-six-foot-tall, fourteen-ton behemoth designed to carry a one-metric-ton warhead over a range of 210 miles. Eventually, it would be renamed the V-2. Building it and making it work required a monumental, multi-year effort, directed by von Braun. While in command of the operation, he and his fellow engineer Rudolf Hermann flew to Berlin to report on the progress. Afterward they stopped for a drink. The barmaid refused to serve the young von Braun until she saw his ID.

In March 1939, von Braun and his crew were called to demonstrate their progress for Hitler himself. They set up two static firings, where the rocket motors were fixed to the ground. From a nearby visitor stand, Hitler watched the two massive engines roar to life. He was visibly unimpressed. Dornberger gave the Führer a briefing of the progress at Peenemunde. He sat through it in stony silence. Realizing that their leader had little understanding of the complexities of rocketry, von Braun and Dornberger gave him a quick primer on propulsion. At last, Hitler showed some interest and asked how long before the A-4 would be ready.

It would take until 1942. After many misfires and more than a few launch-pad explosions, on October 3, 1942, the giant rocket, affectionately nicknamed the "Cucumber," rose smoothly from its pad and rose to the unheard of height of fifty miles, getting closer than any other man-made object to space. Many of the rocket scientists danced and wept with joy. In von Braun's version of the story, he didn't recall if he wept or not, but he did admit to having a few celebratory drinks that night. "I don't know about my getting wet outside, but I got very wet inside," he joked.

Von Braun had tried to remain out of politics, but as his responsibilities grew, so did the pressure on him to join the Nazi Party. Refusing outright would have killed his career, or worse, and would have dashed any hopes of sending his beloved rockets into space. The exact date is unclear, but sometime between 1937 and 1938, Wernher von Braun formally became a Nazi.

To complicate matters further, in 1940, von Braun was "invited" to join the military arm of the Nazi Party, the SS, by none other than Reichsführer Heinrich Himmler. This was ultimately part of Himmler's plan to wrest control of the rocket program from the army. When the SS first offered von Braun a commission as a lieutenant, he tried to decline, saying he was too busy with his work for political activity. They didn't buy the excuse and assured him that being in the SS wouldn't cost him any time at all.

Backed into a corner, von Braun went to Dornberger, and asked what he should do. The colonel advised his young friend that rejecting the offer would be taken as a dangerous sign of disloyalty and that Himmler was not used to taking no for an answer. Eventually, the scientist gave in and was appointed an SS lieutenant. Although he apparently did little more than attend SS monthly meetings, von Braun received regular promotions, eventually rising to the rank of major.

By the summer of 1943, von Braun received orders to rush the A-4 into mass production, so Hitler could begin his planned missile attacks on London. Von Braun knew this was unrealistic. Six out of every ten missiles they test-fired failed, often catastrophically. Months of further development and debugging were needed, but the Führer was not known for his patience. Under ever mounting pressure, the rocketeers worked to meet the goal.

On the night of August 17, 1943, the Allies applied their own pressure. The Royal Air Force (RAF) launched a massive bombing raid on Peenemunde. A force of close to six hundred heavy bombers with fighter escorts hammered the base for over an hour.

By the time it was over, the raid had destroyed many buildings and set others on fire. Von Braun ordered, "Every man out, and save the documents!" Then he rushed into the burning building to save what he could from his office safe.

In the aftermath of the attack, von Braun organized his people and began the impossible seeming job or rebuilding. Debris had to be cleared, production facilities had to be rebuilt and machinery and vehicles had to be repaired. During this period, according to those close to him, von Braun slept no more than three or four hours per night, but under his leadership, after two weeks, they were back in operation.

Further air raids followed, and Hitler responded by ordering that the A-4 production be moved underground. He commanded Himmler to see that the job was done. Himmler in turn put his construction chief in charge, SS General Hans Kammler, who had overseen the building and operation of the Third Reich's concentration camps. Kammler used slave labor to expand existing mining tunnels in the Harz Mountains, turning them into a massive production facility that came to be called Mittelwerk (Central Plant). Once they were done blasting and digging out the mountain, the prisoners were put to work in the production of aircraft, submarines, and of course the rockets, under the most brutal of conditions.

Tens of thousands of prisoners, most of them Russian, Polish, and French POWs were housed underground, forced into cramped and unsanitary conditions. The work was unrelenting, and the beatings by guards and Kapos, the slightly more privileged prisoners, were horrific. The weakest soon died from disease or exposure. Those who resisted were summarily hanged. Some estimates put the number of deaths in excess of twenty thousand.

How much von Braun knew of these conditions and what he could have done about them is still a question today. He visited the production facilities on several occasions, and it would have been difficult to miss the stench of death. There are conflicting reports about whether von Braun personally witnessed any of the beatings. Both Wernher von Braun and his younger brother, Magnus, also a rocket scientist, were seen at Mittelwerk, but the physical resemblance between the two makes eyewitness identification of who saw what difficult.

One account provided by Ernst Stuhlinger, a scientist at Peenemunde, tells of von Braun approaching one of the SS guards to say something about the mistreatment of a prisoner, only to be told he should be careful unless he wanted to wear a striped prisoner's uniform himself. Stuhlinger's account, however, was told decades after the fact and is difficult to verify. The question remains. What can be said is that by 1944 von Braun was in a dangerous situation himself.

Facing increasing pressure to produce the promised weapons, von Braun was summoned in February 1944 to appear before Himmler at the SS headquarters in Hochwald. Years later, von Braun told a friend that in those days if you were summoned to see a high official, you didn't know if you were going to be killed or given a medal. With great trepidation, he went. It turned out that Himmler wanted von Braun's help pushing aside Dornberger, who had since become a general, so the SS could consolidate its control of the rocket program. "I can do a lot more for you, Wernher, than those stuffy Army generals," Himmler told him.

As politely as possible, von Braun turned down the offer. His personal loyalty to Dornberger, who had helped and mentored him since his days as a teenaged rocket hobbyist, won out. The meeting seemed to end amicably, and von Braun left. As he put it, "I could see that he was miffed, but I thought little of it—until I was arrested three weeks later by the Gestapo."

On the night of March 15, 1944, Gestapo agents seized von Braun and two other members of his team. They were taken to a prison in nearby Stattin, and held for two weeks. Eventually, von Braun learned that he had been charged with sabotaging the war effort. He was accused of making remarks at a party indicating he was more interested in going into space than bombing London. Further incriminating him, von Braun was forced to travel a great deal as part of his work. Since he was a pilot, he often kept a small plane, a civilian Messerschmitt, for the purpose. Himmler accused von Braun of keeping the plane gassed up, so he could slip out with the rocket secrets and fly to England.

Von Braun was potentially facing the firing squad, but when the court of inquiry convened to look at the charges, General Dornberger came to his rescue. Dornberger had convinced Hitler's minister of war and arms production, Albert Speer that von Braun was so vital to the rocket program that it would fail without him. Speer intervened, and when Dornberger burst into the court room, he had an order signed by Hitler for von Braun's release. The scientist walked away on a ninety-day probation. Of his prison stay, von Braun later said that when they first put him in the cell his first thought was, "What a wonderful chance to sleep" (Ward, 2005).

Von Braun and his team went back to work, but by the time the A-4 rockets, renamed the V-2 (vengeance weapons) were ready for launch, the Allies had already taken Normandy. On September 8, 1944, three months after the D-day invasion, the first V-2 rockets were fired, striking London's East End. Over the next several months the V-2's would be responsible for 2,742 deaths and 6,467 injuries in England alone. When the victims from the rest of Europe are added, the death toll rises to around 5,000. The closest von Braun came to accepting responsibility for the casualties was in a statement he made at a press conference in 1960: "I have very deep and sincere regrets for the victims of the V-2 rockets, but there were victims on both sides. A war is a war, and when my country is at war, my duty is to help win that war, whether or not I had sympathy for the government, which I did not" (Lang, 1951).

By the beginning of 1945, the Third Reich was in its death throes. Seeing the writing on the wall, von Braun called a meeting of his closest and most trusted team members. He pointed out the obvious, told the team how valuable they were and held out the possibility of ultimately making their goal of space travel come true, but they needed to decide which of the Allies to surrender to. The Red Army was closing in on Peenemunde, but the prospect of falling into Soviet hands didn't look appealing. The British, still reeling from the V-2 attacks would probably not hold them in high regard, and surrendering to the French was never seriously considered.

Von Braun made the case for surrendering to the Americans. In later years, he said he favored the United States because of its intense devotion to individual freedom and human rights. At the time, however, the United States seemed like the country most likely to emerge from the war with the resources and where-withal to go into space, and as von Braun put it, "My country had lost two wars in my rather young lifetime. The next time, I wanted to be on the winning side."

They knew the American forces were south of Peenemunde,

so von Braun used his official SS stationery to forge orders for his group to move to the Harz Mountains. With orders in hand, the entire rocket team, some five thousand employees, along with their equipment, scientific instruments, and crate upon crate of vital documents loaded up every truck, train, or automobile they could find, and moved south. Von Braun even put his SS uniform to good use, helping persuade any guards they met to let them pass.

Traveling only at night to avoid Allied air attacks, they moved out. On the way, von Braun was seriously injured in an auto accident when his driver fell asleep at the wheel. He broke his right arm in several places, but once he had it set and put in a cast, the exodus continued. When they reached their destination, most of the group fanned out into nearby villages. They hid their precious documents in abandoned mines and dynamited the entrances. Once that was complete, von Braun, Magnus, Dornberger, and two dozen of their colleagues proceeded to a resort hotel in the mountains, Haus Ingeburg, to await their chance. For weeks, they waited while events unfolded. As von Braun recalled:

There I was, living royally in a ski hotel on a mountain plateau. There were the French below us to the west, and the Americans to the south. But no one, of course, suspected we were there. So nothing happened. The most momentous events were being broadcast over the radio: Hitler was dead, the war was over, an armistice was signed—and the hotel service was excellent.

Little did von Braun suspect that, all of the Allies were looking for them. He and his fellow scientists were on a list of valuable scientific assets. The Americans, the French, the British, and the Russians were all racing each other to find them. This mad scramble was called by Winston Churchill the "Wizard War" (Churchill, 1953).

On May 2, 1945, Magnus von Braun dressed in civilian clothes and rode a bicycle down the mountain in search of some Americans to surrender to. He eventually found an advance anti-tank patrol unit of the 44th Infantry. When one of the soldiers, Private First Class Fred P. Schneiker stopped him, Magnus told the GI in his best English, "We are a group of rocket specialists up in the mountains. We want to see your commander and surrender to the Americans." He also added that the group wanted to be, "taken to see 'Ike' as soon as possible." The private thought he was nuts.

When the soldiers finally took Magnus to Counter-Intelligence Corps headquarters, the officer in charge didn't know anything about the search for the German scientists. He simply told Magnus to come back the next morning with the leaders of the group. Shortly after dawn the next day, a convoy of cars bearing the von Braun brothers and some of the most valuable scientists in the world showed up.

Von Braun talked quite freely with the American GIs and posed for pictures with them. He bragged of his achievements in rocketry and talked of going into space one day. After a few days of this, one of the soldiers remarked that they had captured "either the biggest scientist in the Third Reich or the biggest liar!"

A lengthy interrogation ensued while the Americans rushed to round up the rest of the rocket team and find the literally tons of documents that they had concealed. Once that was done, a U.S. official reported, "One of the greatest scientific and technical treasures in history is now securely in American hands." When Stalin learned that he had been beaten to the prize, he roared, "We defeated the Nazi armies; we occupied Berlin and Peenemunde; but the Americans got the rocket engineers!"

Once the value of the German scientists was fully recognized, it was decided to move the most important ones to the United States. Von Braun was asked to make recommendations. His original list had over five hundred names on it. The army balked, and eventually the list was whittled down to 127 crucial individuals. As the file of each prospective candidate was reviewed, those that were selected were marked with a paperclip. The operation thus came to be known as Project Paperclip.

Not wanting to upset the American public with news that Nazi scientists were being brought to U.S. soil, the military flew von Braun and the first group of them in secretly in September 1945. At first they were held at Fort Strong in Boston harbor, but eventually, ended up at Fort Bliss, Texas, outside El Paso. Von Braun and his fellow Germans ended up spending the next several years there, sharing their knowledge with the Americans and teaching them about the captured V-2s.

When not working on the rockets, von Braun and the others spent their time learning English. Amusingly, since they were learning mainly from those in and around the fort, what they picked up was English with a Texas twang. What's worse they had picked up American idioms and slang from the GIs. When von Braun gave his first presentation for officials in Washington, he didn't understand why they kept laughing at certain parts of his talk. When he was done, one of his handlers, Major James Hamill took him aside and told him, "Wernher, before you give your next talk, let me go over your manuscript. You just cannot use those GIs' slang expressions with decent people!" (Bergaust, 1976).

Von Braun returned to Germany briefly in 1947. He had become engaged to one of his cousins, Maria Luise von Quistorp, and the army gave permission for him to fly home and marry her. While in Germany, he was kept under tight military security, to prevent kidnap attempts by the Soviets. Apparently, that put a bit of a crimp in the honeymoon, since the military policemen insisted on staying in the apartment he and his new bride were occupying. After a few days, he returned to Fort Bliss with his new wife and his elderly parents. It took another year before von Braun and his wife would get a proper honeymoon.

Back in America, von Braun worked hard to keep his dream of space alive. Shortly after the military revealed he and the other Germans were in the country, von Braun began speaking publicly about space exploration to any group that would listen. One of his first talks was to the El Paso Rotary Club. After addressing the military uses of rockets, he began telling the audience about the future of going into space, satellites, orbiting space stations, exploring other worlds, and the wonders of floating weightlessly. When he was done, they gave him a standing ovation.

While working on missile projects for the military, von Braun continued his one-man campaign for space. In addition to his public speaking, he began working on a science fiction novel to popularize his ideas. It was called *Mars Project* and included extensive technical appendixes in addition to the story. Von Braun wasn't able to find an American publisher, but eventually, it was printed in Germany. His efforts were rewarded in 1949, when the British Interplanetary Society invited him to become an honorary fellow, "in recognition of your great pioneering activities in the field of rocket engineering." Von Braun was deeply touched by this gesture by his former enemies.

In 1950, the military moved von Braun and the other Germans to Huntsville, Alabama, to work on the Redstone Missile project. Von Braun continued to display the leadership and managerial skills that he had perfected at Peenemunde. He and his fellow Germans found the green hills of Huntsville much closer to their homeland than the barren desert of Fort Bliss. They settled down, becoming part of the community. Von Braun built a home there and began raising a family. Eventually, he and Maria would have three children, two girls and a boy. In 1955, Wernher von Braun, along with forty members of his team, became a U.S. citizen.

Von Braun's campaign for space exploration got a big boost in 1951, when Collier's magazine decided to do a series of articles on it. Their interest had been piqued by a symposium at New York's Hayden Planetarium. Once the magazine's writer/editor Connie Ryan met von Braun and sat down with him over drinks, they were hooked. Between 1952 and 1954, Collier's ran eight articles written by von Braun and others, making a well-reasoned and impassioned case for going into space. They were accompanied by stunning illustrations by Chesley Bonestell, including one showing a now familiar wheel-shaped space station of von Braun's design.

The articles were a hit, and the public's fascination was furthered by a series of space-centered episodes of the new Walt Disney television program. The episodes, which von Braun worked on, began airing in 1955. The response from the audience was overwhelmingly positive. Less than a year after Robert Oppenheimer, the American hero and father of the atomic bomb, was pilloried in the press by the anticommunist witch hunters, Wernher von Braun, the father of the V-2 rocket, was the most popular scientist in the country.

Within a few years, von Braun and the army's Redstone rocket program were starting to see some real success. Coincidentally, 1957–1958 had been declared the International Geophysical Year, an international research initiative emphasizing Earth and its atmosphere. Von Braun and others used this as an opportunity to propose using their Redstone rocket to launch a satellite into orbit. He got the go ahead from the army, but what ensued was an interservice power struggle. The army, the navy and the air force were all competing for funding and influence in Washington.

Instead of going with the army's tried-and-true Redstone, the Eisenhower administration put its support behind Naval Research Laboratory's Vanguard program. The Vanguard was untested, but the administration declared its design more "dignified" than the army's Redstone. Von Braun responded, "I'm all for dignity, but this is a Cold War tool. How dignified would our position really be if a man-made star of unknown origin suddenly appeared in our skies?"

Von Braun tried to point out flaws in the Vanguard design but to no avail. Even though he was confident his Redstone rocket was capable of reaching space, he was ordered not to proceed. The Vanguard ran into problem after problem, much as von Braun had predicted. On October 4, 1957, before the navy's Vanguard could be launched, the Soviet Union put a 184-pound, spherical satellite named *Sputnik* (Little Moon) into orbit. The Russians had beaten the Americans being the first into space. The feeling of technological superiority that America had enjoyed in the postwar years was shattered.

Von Braun was in the Redstone Arsenal Officers Club with a roomful of visiting Defense Department and army brass when the news broke. He exclaimed, "I'll be damned," and then launched into an appeal to the Washington visitors. "We knew they were going to do it! Vanguard will never make it! We have the hardware on the shelf! For God's sake, turn us loose and let us do something!" von Braun pleaded. It was no use. The Vanguard program continued. On December 6, 1957, the Vanguard was on the launch pad, ready to take America's satellite into orbit. After the countdown concluded, the command was given for lift-off. In front of television cameras from around the nation, it exploded spectacularly. The press labeled it "Kaputnik."

Without bothering to gloat, von Braun and his team went to work. On January 31, 1958, less than sixty days from the Vanguard disaster, the army's Number 29 super-Redstone (later renamed the Jupiter-C) was ready. It blasted off at 10:55 P.M. EST. Von Braun had calculated they should start receiving radio signals from the satellite, named *Explorer*, by 12:41, 106 minutes after the initial lift-off. When the appointed time came and went, nothing happened. Von Braun and the others sweated bullets waiting for some response. At last, after eight minutes, at 12:49, the tracking stations began receiving signals from the orbiting satellite. America was officially in space.

Now, the race was on. The Soviets had followed up on their initial victory by launching *Sputnik II*, on November 3, 1957. It carried the first animal into space, as the dog Laika went into orbit. Eisenhower responded by sending Congress a proposal to establish a national space agency. In mid-July, a few months after the Russians put *Sputnik III*, a massive three-thousand-pound satellite, into orbit, Congress authorized the formation of the National Aeronautic and Space Administration (NASA). Von Braun and his team were a central part of the new organization.

By 1959, the U.S. space program had managed to put its own animals in space, a rhesus monkey named Able and a squirrel monkey named Baker. Unlike Laika, the monkeys were returned safely to Earth. At that point, work began in earnest on the next great challenge, putting a human in space. The American effort was called Mercury, after the messenger of the gods.

Von Braun and his team worked hard on developing the Saturn rockets to carry the first astronauts into space at the newly christened Marshall Space Flight Center, in Huntsville. Although much of the training of the first seven astronauts took place at other facilities, von Braun enjoyed spending as much time as possible with them. When asked at a news conference if he would like to go into space with them, von Braun said, "Sure, why not? I envy them. But they just told me I'm too fat!"

Preparations for the flight were almost completed in the spring of 1961, when the Russians once again beat the Americans to the punch. On April 12, Soviet Major Yuri Gagarin became the first human being in space. He completed an orbit of the planet and was returned safely to Earth. It wouldn't be until May 5, that the first American, Alan Shepard, would venture into space. Von Braun was again disappointed when he heard of Gagarin's success, and he told the press, "We are going to have to run like Hell to catch up!"

President John F. Kennedy tried to take back the initiative. He went to Vice President Lyndon Johnson and asked him to recommend a national space project, something dramatic that the

United States could use to win the competition for space with the Russians. Johnson went to von Braun, and a number of others, and asked for input. Von Braun responded with a ten-page letter proposing a manned landing on the Moon. He laid out the benefits of such an accomplishment and declared the United States stood an excellent chance of being able to accomplish it by 1970. Johnson read the recommendation and gave a formal report to the president. On May 25, 1961, before a joint session of Congress, President Kennedy challenged the nation to, "commit itself to achieving the goal before this decade is out, of landing a man on the Moon and returning him safely to Earth."

Many were skeptical that the new Apollo project could succeed, and some, including former President Eisenhower, called it a "stunt." Von Braun pointed out that many had called Charles Lindbergh's 1927 solo flight across the Atlantic a stunt as well. He assured the doubters that Apollo would be, "the wisest investment America has ever made."

On July 21, 1961, Major Virgil "Gus" Grissom became the second American in space. That was followed up on February 20 of 1962 with Lieutenant Colonel John Glenn's successful completion of the U.S.'s first orbital mission. The following year, in May1963, Kennedy went to the Marshall Space Flight Center to check the progress for himself. Von Braun treated the president to a demonstration firing of one of the Saturn boosters. From the safety of an outdoor observation bunker, Kennedy witnessed the fiery spectacle of the huge rocket shake the ground it was tethered to. When it was done, the president grabbed von Braun's hand, and said, "That's just wonderful! . . . If I could only show all this to the people in Congress!"

Six months later, President John F. Kennedy was assassinated in Dallas. Von Braun, like the rest of the nation, was struck deeply by the loss. Later, the German scientist wrote a personal letter of condolence to the president's wife. At the end of the letter, he wrote:

> You have been overwhelmed with condolences from all over the world at the tragic death of your beloved husband. Like for so many, the sad news from Dallas was a terrible personal blow to me. We do not know a better way of honoring the late President than to do our very best to make his dream and determination come true that "America must learn to sail on the new ocean of space, and be in a position second to none."

A few days later, von Braun received a handwritten response from Mrs. Kennedy including the following: "What a wonderful world it was for a few years—with men like you to help realize his dreams for this country—And you with a President who admired and understood you—so that together you changed the way the world looked at America—and made us proud again" (Ward, 2005).

The work continued. When the Saturn V that would carry the astronauts to the moon was completed, it towered 365 feet. Including the Apollo space craft, it was a full six stories taller than the Statue of Liberty. It weighed 6.5 million pounds and contained over three million parts. Together its five engines were capable of delivering 7.5 million pounds of thrust. That's equivalent to the combined power of eighty-five Hoover Dams. NASA Administrator Keith T. Glennan called it, "one of the most amazing combinations of engineering, plumbing, and plain hope anyone could imagine" (Wright, 1998).

While the Marshall Center team was working feverishly to finish their mighty rocket, von Braun kept his eye on the future. There were those, however, who would not let him forget the past. In early 1964 Stanley Kubrick released his motion picture parody *Dr. Strangelove, or: How I learned to Stop Worrying and Love the Bomb*. While the main targets of the film were nuclear proliferation and the cold war, the title character, Dr. Strangelove was an ex-Nazi mad scientist who bore an uncomfortable resemblance to von Braun. According to most reviewers, the character was an amalgam of von Braun, Edward Teller, and Herman Kahn, but at the film's climactic moment Strangelove stands up from his wheelchair and says, "Mein Führer, I can walk!" Few could mistake who it was aimed at.

It was followed the next year by an even less subtle bit of humor. Tom Lehrer, a Harvard and MIT mathematician, who parlayed his musical talent and sharp wit into a second career as a satirist, wrote a song titled "Wernher von Braun." Released as part of Lehrer's 1965 album, That Was the Year That Was, it lyrically portrayed the German rocket scientist as an opportunist, willing to work for anyone who would pay for his services. The song concluded with a none too subtle tribute to the Doctor's linguistic talent. It acknowledged his ability to count down to zero in German or English, and hinted at his plans to learn Chinese.

Eventually, some of the controversy died down, but a few years later on January 27, 1967, tragedy struck. The Apollo 1 astronauts, Gus Grissom, Edward H. White, and Roger B. Chaffee were in the command capsule at Cape Kennedy (formerly Cape Canaveral) for a routine checkout. As they sat in the capsule, an electrical short-circuit started a fire. The pure oxygen in the capsule caused the fire to spread rapidly. All three astronauts died.

Shortly after von Braun got the news, he spoke to the press, and described the three lost men as, "three good friends and valiant pioneers." He continued, "Their deaths brought to mind the Roman saying 'per aspera ad astra'—'a rough road leads to the stars.'"

Again, the work continued. Redesigns were implemented to prevent a repeat of the Apollo 1 accident, and Apollo missions 2 through 6 were unmanned. By October 1968, the team was sufficiently confident to launch Apollo 7 into Earth orbit with astronauts Wally Schirra, Walter Cunningham, and Donn Eisele aboard. It was a success, and was quickly followed by a Christmas launch of Apollo 8. Frank Borman, Jim Lovell, and Bill Anders became the first humans to orbit the Moon.

Apollo 9 took off in March 1969 to test the command service and lunar modules in Earth orbit. Next was Apollo 10. Launched in May, it completed a fly-by of the Moon, coming tantalizingly close to its surface. In July it came time for the launch of Apollo 11. The dream that von Braun had nurtured since he was a boy in Germany, that he had worked and struggled for, was at hand. Two days before the launch, von Braun was asked by a reporter to rank the impending lunar landing with other events in history. He responded, "About with the importance of aquatic life first crawling on land."

On the morning of the historic launch, von Braun arrived at the control center at Cape Kennedy at 4:00 A.M. After hours of nerve-wracking checks and rechecks, the massive Saturn V rocket bearing astronauts Neil Armstrong, Buzz Aldrin, and Michael Collins was ready. It blasted off at 9:32 A.M. July 17. With a reported one million spectators present and millions more watching on television, it rose majestically into the air, dropping spent stages as it sped to its parking orbit around Earth. Once there, von Braun's baby sent the three courageous Americans on the three-day, quarter of a million mile journey to the Moon.

On July 19, the Apollo 11 spacecraft reached its lunar orbit. The next day, with Collins keeping the command vehicle in orbit, Armstrong and Aldrin piloted the lunar lander to the Moon's cratered surface. With von Braun and the others waiting anxiously back on Earth, they touched down, and Armstrong delivered the message, "Tranquility Base here. The Eagle has landed." A few hours later, he emerged from the lander, and climbed down the ladder, uttering the historic words, "That's one small step for a man, a giant leap for mankind." Von Braun's dream had come true.

The Apollo 11 Moon landing marked the high point of von Braun's life and the culmination of his career. However, even before Armstrong, Aldrin, and Collins returned to Earth, the dismantling of the manned space program had begun. In the midst of the Vietnam War and the social upheavals of the 1960s and early 1970s, many argued, "we shouldn't be wasting our money in space when we can put it to better uses here on Earth." NASA lost funding and valuable personnel, and the last manned mission to the Moon, Apollo 17, ended on December 1972. Even though the space program continued, with the launching of *Skylab,* the International Space Station and the Hubble space telescope, for the time being the dream of humans setting foot on another world was on hold.

Von Braun always envisioned that space exploration would have benefits for humanity far beyond bringing back a few handfuls of lunar dirt. He was right. Telecommunication satellites, GPS systems, desk sized computers, and many other advances all owe their origins ultimately to von Braun's work. His entire life was dedicated to the realization of his vision, a vision he summed up in a speech he delivered shortly after Explorer I reached orbit:

> We have stepped into a new, high road from which there can be no turning back. As we probe farther into the area beyond our sensible atmosphere, man will learn more about his environment; he will understand better the order and beauty of creation. He may then come to realize that war, as we know it, will avail him nothing but catastrophe. He may grasp the truth that there is something much bigger than his one little world. Before the majesty of what he will find out there, he must stand in reverential awe. This then, is the acid test as man moves into the unknown.

BIBLIOGRAPHY OF WORKS CITED

Royal Society Journal Book Copy (1767–1770).

Ackroyd, Peter. *Ackroyd's Brief Lives: Newton.* New York: Doubleday, Nan A. Talese, 2006.

Ball, Philip. *The Devil's Doctor.* New York: Farrar, Straus and Giroux, 2006.

Bergaust, Erik. *Whernher von Braun.* Washington, D.C.: National Space Institute, 1976.

Bird Kai, and Martin J. Sherwin. *American Prometheus: The Triumph and Tragedy of J. Robert Oppenheimer.* New York: Alfred A. Knopf, 2005.

Braun, Wernher von. "Space Man: The Story of My Life." American Weekly, (July 20, 1958): 8.

———. "Speech before the National Military-Industrial Conference, Chicago." Chemical & Engineering News, (March 3, 1958): 56.

Carter, K. Codell. "Ignaz Semmelweis, Carl Mayrhofer, and the Rise of Germ Theory." *Medical History* (1985): 33–53.

Christianson, Gale E. *Isaac Newton.* New York: Oxford University Press, 2005.

Churchill, Winston S. *Triumph and Tragedy.* Boston: Houghton Mifflin, 1953.

Clark, Ronald. *Einstein: The Life and Times.* New York: Harper Collins, 1971.

Cornelius, and A. J. Harding Rains, eds. *Letters from the Past, from John Hunter to Edward Jenner.* London: Royal College of Surgeons of England, 1976.

Curie, Eve. *Madame Curie: A Biography by Eve Curie.* Garden City, NJ: Doubleday, Doran, 1937.

Curie, Marie. *Autobiographical Notes*. New York: MacMilllan, 1923.

———. *Pierre Curie*. New York: MacMillan, 1923.

Davy, Humphry. *Researches, Chemical and Philosophical, Chiefly Concerning Nitrous Oxide and its Respiration*. London: Johnson, 1799.

———. *The Collected Works of Sir Humphry Davy*. London: Smith, Elder, 1839–1840.

Davy, John. *Memoirs of the Life of Sir Humphry Davy*. London: Churchill, 1858.

Dobson, Jessie. "Some of John Hunter's Patients." *Annals of the Royal College of Surgeons of England*, (1968): 124–133.

Donovan, Sandy. *Hypatia: Mathematician, Inventor, and Philosopher*. Minneapolis: Compass Point Books, 2008.

Einstein, Albert. "What I Believe." *Forum and Century*, (1930): 194.

Goldsmith, Barbara. *Obsessive Genius: The Inner World of Marie Curie*. New York: W.W. Norton & Company, Inc., 2005.

Hirshfeld, Alan. *Eureka Man*. New York: Walker Publishing Company, Inc., 2009.

Hoffmann, Banesh. *Albert Einstein: Creator and Rebel*. New York: Viking, 1972.

Isaacson, Walter. *Einstein*. New York: Simon & Schuster, 2007.

———. *Einstein: His Life and Universe*. New York: Simon & Schuster, 2007.

Jackson, Joe. *A World on Fire*. New York: Penguin Books, 2005.

Johnson, Steven. *The Invention of Air*. New York: Riverhead Books, 2008.

Kaku, Michio. *Einstein's Cosmos*. New York: W.W. Norton & Company, Inc., 2004.

Knight, David. *Humphry Davy: Science and Power*. Cambridge: Cambridge University Press, 1992.

Lang, Daniel. "A Romantic Urge." *New Yorker*, (April 21, 1951): 75–93.

Magyar, Laszlo A. "John Hunter and John Dolittle." *Journal of Medical Humanities* (1994): 217–220.

Moore, Wendy. *The Knife Man*. New York: Broadway Books, 2005.

Neufeld, Michael J. *Von Braun: Dreamer of Space, Engineer of War*. New York: Alfred A. Knopf, 2007.

Neufeld, Michael. *The Rocket and the Reich: Peenemunde and the Coming of the Ballistic Missile Era.* New York: Simon & Schuster, Free Press, 1995.

Nuland, Sherwin B. *The Doctors' Plague: Germs, Childbed Fever, and the Strange Story of Ignác Semmelweis.* New York: W. W. Norton & Company, Inc., 2003.

O'Neill, John J. *Prodigal Genius.* New York: David McKay Co., 1944.

Paget, Stephen. *John Hunter: Man of Science and Surgeon.* London: Fischer Unwin, 1897.

Palmer, James, ed. *The Works of John Hunter.* London: Longman, Rees, Orme, Brown, Breen, 1835.

Parent, André. "Giovanni Aldini: From Animal Electricity." *The Canadian Journal of Neurological Sciences* (2004): 576–84.

Pollard, Justin, and Howard Reid. *The Rise and Fall of Alexandria: Birthplace of the Modern Mind.* New York: Penguin, 2006.

Priestley, Joseph. *Experiments and Observations on Different Kinds of Air and Other Branches of Natural Philosophy.* Birmingham, England: Thomas Pearson, 1790.

———. "Letter to Franklin." July 1, 1772.

"Reach for the Stars." *Time,* (February 17, 1958): 24.

Schofield, Robert E. *A Scientific Autobiography of Joseph Priestley.* Cambridge: MIT Press, 1966.

Shelley, Mary. *Frankenstein.* 1819.

Smith, Adam. *The Correspondence of Adam Smith.* 1787.

Solomon, Joan. *Structure of Matter: The Growth of Man's Ideas on the Nature of Matter.* Newton Abbot: David and Charles, 1973.

Terneer, Anne. *Mercurial Chemist: A Life of Sir Humphry Davy.* London: Methuen, 1963.

Tesla, Nikola. "My Inventions." *Electrical Experimenter* (1919): 30.

The Catholic Encyclopedia. New York: Robert Appleton Company, 1907.

U.S. Department of Energy. The Manhattan Project: An Interactive History. April 22, 2010 <http://www.cfo.doe.gov/me70/manhattan/hiroshima.htm>.

Vrettos, Theodore. *Alexandria: City of the Western Mind.* New York: The Free Press, 2001.

Ward, Bob. *Dr. Space: The Life of Wernher von Braun.* Annapolis: Naval Institute Press, 2005.

Westfall, Richard S. *Short-Writing and the State of Newton's Conscience, 1662.*

Wright, Mike. "Huntsville and the Space Program." *Alabama Heritage* spring and summer 1998: 27.